Human Development and Global Institutions

This book provides a timely and accessible introduction to the foundational ideas associated with human development. It examines the concept's evolution during the post-colonial era, and discusses how various institutions of the UN system have tried to engage with this issue, both in terms of intellectual and technical advances, and operationally. Showing that human development has had a profound impact on shaping the policy agenda and programming priorities of global institutions, Ponzio and Ghosh argue that human development has helped to preserve the continued vitality of major multilateral development programs, funds, and agencies.

It also details how human development faces new risks and threats, caused by political, economic, social, and environmental forces, that are highlighted in a series of case studies on trade, water, energy, the environment, democracy, human rights, and peacebuilding. The book also makes the case for why human development remains relevant in an increasingly globalized world, asking whether global institutions will be able to sustain political and moral support from their member states and powerful nonstate actors. The authors argue that fresh new perspectives on human development are now urgently needed to fill gaps across borders and entire regions. A positive, forward-looking agenda for the future of global governance would have to engage with new issues such as the Sustainable Development Goals, energy transitions, resource scarcity, and expansion of democratic governance within and between nations.

By redefining what constitutes human progress in an increasingly interdependent world, this book serves as a primer for scholars and graduate students of international relations and development. It is also relevant to scholars of economics, political science, history, sociology, and women's studies.

Richard Ponzio is head of the Global Governance Program at The Hague Institute for Global Justice, where he serves as project director for the Commission on Global Security, Justice & Governance. He is formerly a senior adviser in the US State Department's Office of the Special Representative for Afghanistan and Pakistan. Richard has served with the UN in Afghanistan, Bosnia, Kosovo, New York, Pakistan, Sierra Leone, and the Solomon Islands.

Arunabha Ghosh is CEO, Council on Energy, Environment and Water, India. Widely published, with work experience in 35 countries and having previously worked at Princeton, Oxford, the UN Development Programme, and the World Trade Organization, Arunabha advises governments, industry, and civil society around the world on: energy and resources security, renewable energy, water governance, climate governance, energy-trade-climate linkages, and international regimes. He is, most recently, co-author of *Climate Change: A Risk Assessment* (2015).

Global Institutions

Edited by Thomas G. Weiss
The CUNY Graduate Center, New York, USA
and Rorden Wilkinson
University of Sussex, Brighton, UK

About the series

The "Global Institutions Series" provides cutting-edge books about many aspects of what we know as "global governance." It emerges from our shared frustrations with the state of available knowledge—electronic and print-wise, for research and teaching—in the area. The series is designed as a resource for those interested in exploring issues of international organization and global governance. And since the first volumes appeared in 2005, we have taken significant strides toward filling conceptual gaps.

The series consists of three related "streams" distinguished by their blue, red, and green covers. The blue volumes, comprising the majority of the books in the series, provide user-friendly and short (usually no more than 50,000 words) but authoritative guides to major global and regional organizations, as well as key issues in the global governance of security, the environment, human rights, poverty, and humanitarian action among others. The books with red covers are designed to present original research and serve as extended and more specialized treatments of issues pertinent for advancing understanding about global governance. And the volumes with green covers—the most recent departure in the series—are comprehensive and accessible accounts of the major theoretical approaches to global governance and international organization.

The books in each of the streams are written by experts in the field, ranging from the most senior and respected authors to first-rate scholars at the beginning of their careers. In combination, the three components of the series—blue, red, and green—serve as key resources for faculty, students, and practitioners alike. The works in the blue and green streams have value as core and complementary readings in courses on, among other things, international organization, global governance, international law, international relations, and international political economy; the red volumes allow further reflection and investigation in these and related areas.

The books in the series also provide a segue to the foundation volume that offers the most comprehensive textbook treatment available dealing with all the major issues, approaches, institutions, and actors in contemporary global governance—our edited work *International Organization and Global Governance* (2014)—a volume to which many of the authors in the series have contributed essays.

Understanding global governance—past, present, and future—is far from a finished journey. The books in this series nonetheless represent significant steps toward a better way of conceiving contemporary problems and issues as well as, hopefully, doing something to improve world order. We value the feedback from our readers and their role in helping shape the on-going development of the series.

A complete list of titles appears at the end of this book. The most recent titles in the series are:

Human Development and Global Institutions (2016)
by Richard Ponzio and Arunabha Ghosh

NGOs and Global Trade (2016)
by Erin Hannah

Brazil as a Rising Power (2016)
edited by Kai Michael Kenkel and Philip Cunliffe

Summits and Regional Governance (2015)
edited by Gordon Mace, Jean-Philippe Thérien, Diana Tussie, and Olivier Dabène

Global Consumer Organizations (2015)
by Karsten Ronit

World Trade Organization (2nd edition, 2015)
by Bernard M. Hoekman and Petros C. Mavroidis

Women and Girls Rising (2015)
by Ellen Chesler and Terry McGovern

Human Development and Global Institutions

Evolution, impact, reform

Richard Ponzio and Arunabha Ghosh

Routledge
Taylor & Francis Group
LONDON AND NEW YORK

First published 2016
by Routledge
2 Park Square, Milton Park, Abingdon, Oxon OX14 4RN

and by Routledge
605 Third Avenue, New York, NY 10017

Routledge is an imprint of the Taylor & Francis Group, an informa business

British Library Cataloguing in Publication Data
A catalogue record for this book is available from the British Library

Library of Congress Cataloging in Publication Data
Names: Ponzio, Richard, author. | Ghosh, Arunabha, author.
Title: Human development and global institutions : evolution, impact, reform / Richard Ponzio and Arunabha Ghosh.
Description: Abingdon, Oxon ; New York, NY : Routledge, 2016. | Series: Routledge global institutions series | Includes bibliographical references and index.
Identifiers: LCCN 2015040270| ISBN 9780415483599 (hardback) | ISBN 9780415483605 (pbk.) | ISBN 9781315640945 (ebook)
Subjects: LCSH: Human security–International cooperation. | Human services–International cooperation. | Social service–International cooperation. | Community development–International cooperation. | Economic development–International cooperation. | United Nations.
Classification: LCC JC571 .P656 2016 | DDC 303.44–dc23
LC record available at http://lccn.loc.gov/2015040270

ISBN: 978-0-415-48359-9 (hbk)
ISBN: 978-0-415-48360-5 (pbk)
ISBN: 978-1-3156-4094-5 (ebk)

Typeset in Times New Roman
by Taylor & Francis Books

We dedicate this book to our parents—Rabindra Kumar Ghosh and Geetika Ghosh, and Carol Lee Ponzio and Richard Joseph Ponzio—for always enlarging our choices and expanding our freedoms

Contents

List of illustrations

Figures

Table

Acknowledgments

This project originates from our personal engagement with the "human development school of thought" since early, formative years in our professional lives. We feel privileged to have served in the Human Development Report Office of the UN Development Programme (Richard and Arunabha, at different times, under the leadership of Sakiko Fukuda-Parr, and Arunabha under the subsequent leadership of Kevin Watkins). Richard also served as a Visiting Fulbright Fellow under Mahbub and Khadija Haq at their Human Development Centre in Islamabad (later renamed the Mahbub ul Haq Human Development Centre following Dr. Haq's sudden passing in 1998). As the chief architect of the human development paradigm (along with Amartya Sen), Mahbub always held dear the promise of global institutions to deliver on fundamental human development priorities. We are thankful to these intellectual leaders and numerous other colleagues who have inspired and informed our thinking on the rich, multifaceted concept of human development, which continues to shape international policy debates and the direction of global institutions well into its third decade.

We also wish to express our gratitude to Routledge, the project's publisher, and, in particular, its *Global Institutions Series* editors Thomas Weiss and Rorden Wilkinson for their intellectual guidance, moral support, good humor, and remarkable patience. We wish to thank Martin Burke for his expert copy-editing and formatting assistance. We are further thankful for the support and substantive feedback received from our colleagues at The Hague Institute for Global Justice and the Council on Energy, Environment and Water, as well as participants in the Annual Meeting of the Academic Council on the United Nations System held in June 2014 at Kadir Has University in Istanbul. We are grateful to the Rockefeller Foundation Bellagio Center, which hosted Arunabha as a Resident Fellow in August–September 2013, and where some of the ideas and arguments presented here were researched and refined. We wish to extend a special thanks for the intellectual insights and encouragement provided by George Gray Molina.

Finally, we wish to express our appreciation to our families, particularly our wives Kuniko and Meghana, for their steadfast support since the inception of this project. The birth of Arunabha's daughter, Nandita, during this project might have delayed the writing by a few months but brought home the core value of human development: the enlargement of choices and freedoms for everyone, including our next generation!

Abbreviations

AIDS	Acquired immune deficiency syndrome
APEC	Asia-Pacific Economic Cooperation
BASIC	Brazil, South Africa, India, and China (coalition in climate change negotiations)
BRICS	Brazil, Russia, India, China, and South Africa
CCAC	Climate and Clean Air Coalition (on short-lived climate pollutants)
COIN	counter-insurgency operations
DAC	Development Assistance Committee
DPA	UN Department of Political Affairs
ECOSOC	UN Economic and Social Council
EI	Education Index
ESC	Economic Security Council
G8	Group of Eight
G20	Group of Twenty
GATT	General Agreement on Tariffs and Trade
GDP	Gross Domestic Product
GDI	Gender-related Development Index
GEF	Global Environment Facility
GEM	Gender Empowerment Measure
GII	Gender Inequality Index
GNP	Gross National Product
HDI	Human Development Index
HDN	Human Development Network
HDR	*Human Development Report*
HIV	Human immunodeficiency virus
HPI	Human Poverty Index
IAOS	International Association for Official Statistics
ICC	International Criminal Court
IGO	Intergovernmental organization
IHDI	Inequality-adjusted Human Development Index

IMF	International Monetary Fund
IPC	Intellectual Property Committee
IPF	Indo-Pacific Forum
IPR	Intellectual property rights
ITO	International Trade Organization
LEI	Life Expectancy Index
MDG	Millennium Development Goal
MPI	Multidimensional Poverty Index
MSF	Médecins Sans Frontières
MYSI	Mean Years of Schooling Index
NATO	North Atlantic Treaty Organization
NGO	Nongovernmental organization
OAS	Organization of American States
ODA	Overseas development assistance
OECD	Organisation for Economic Co-operation and Development
OSCE	Organization for Security and Co-operation in Europe
PBC	UN Peacebuilding Commission
PBF	UN Peacebuilding Fund
PBSO	UN Peacebuilding Support Office
PFI	Political Freedom Index
PPP	Purchasing power parity
PRSP	*Poverty Reduction Strategy Paper*
R2P	Responsibility to Protect
R&D	Research and development
SDG	Sustainable Development Goal
UN	United Nations
UNCTAD	UN Conference on Trade and Development
UNDAF	UN Development Assistance Framework
UNDP	United Nations Development Programme
UNESCO	UN Educational, Scientific and Cultural Organization
UNFCCC	UN Framework Convention on Climate Change
UNGA	UN General Assembly
UNHCR	UN High Commissioner for Refugees
UNICEF	United Nations Children's Fund
UNMIK	UN Mission in Kosovo
UNWCED	United Nations World Commission on Environment and Development
USAID	United States Agency for International Development
WDR	World Development Report
WHO	World Health Organization
WIPO	World Intellectual Property Organization
WTO	World Trade Organization

Introduction

- **The book's purpose and intended audience**
- **Methodology and structure**

In a world where scholars and policy-makers alike still debate whether we have a consensus on what development is and how it should be done, the human development paradigm—focused, in short, on "the process of enlarging people's choices"—has represented, in a relatively short period of time beginning in the 1990s, the closest the world has come to forging such a global consensus. This is particularly the case in international policy circles where human development priorities served, arguably, as forerunners to the highly acclaimed Millennium Development Goals (2000–2015) and their "Post-2015 Development Agenda" successors, the Sustainable Development Goals (2015–2030) unveiled at the UN Sustainable Development Summit in September 2015 in New York.

Even many of human development's detractors—several of whom are referenced in this volume—concede that this new school of thought, formally introduced globally in the first *Human Development Report* (*HDR*) in 1990, has played a pivotal role in shoring up agreement, at the very least, that governments and their international partners should continue to invest in development, even as the strategic interests of great powers shifted and receded following the 40-year Cold War.[1]

Similar to its influence on the theory, policy, and practice of development over the past quarter century, human development has had a profound impact on shaping the policy agenda and programming priorities of global institutions, the subject of this book. In addition to redefining the overall nature and specific characteristics of what constitutes human progress in an increasingly integrated and interdependent world, human development has helped, for instance, to preserve the continued vitality, and in some cases basic survival, of major multilateral development programs, funds, and agencies almost entirely dependent on donor countries. At the same time, human development, as well as wider policy and scholarly interest in development, faces new risks and threats today—caused by political, economic, social, and environmental forces detailed in this volume.

These risks and threats are intertwined with whether global institutions too will be able to sustain political and moral support from their member states, as well as from a growing number of powerful nonstate actors from within civil society and the business community. Fresh new perspectives on human development, which inform practical institutional reforms, are now urgently needed to fill critical gaps across borders and entire regions, as well as to provide a positive, forward-looking agenda for the future of global governance.

Below, the introduction summarizes the book's purpose and main intended audience, as well as how it complements other publications on human development. It concludes by outlining the study's methodology, structure, and the focus of the subsequent chapters.

The book's purpose and intended audience

Earlier titles have traced the evolution of human development in a variety of sectors and levels of governance.[2] They have, for example, explored it conceptually, in advancing new measurement tools, and in contributing critical analysis and innovative policy prescriptions. In contrast, this volume specifically examines human development's influence in redefining human progress over the past two-and-a-half decades in institutions of global governance. It also makes the case for why human development remains relevant in an increasingly globalized world.

In considering the degree to which human development priorities have shaped policy discussions in global institutions, and the outcomes engendered in connection with these deliberations, two important target groups for this book are policy-makers and scholars. The book also serves as a primer for university-level students of international relations and development seeking to learn about whether and how international policy-making about human development has improved living conditions for the vast majority of people, including the poor and vulnerable. In this regard, it also aims to benefit students as well as scholars of economics, political science, history, sociology, and women's studies in both developed and developing countries.

This study provides a timely and accessible introduction to the foundational ideas associated with the human development school of thought. In addition to the hundreds of global, regional, national, and sub-national *HDR*s produced, the human development school of thought has evolved and grown through a *Journal of Human Development and Capabilities*, the Global Forum on Human Development, the Human Development and Capability Association with over 1,000 members, two global e-networks (HDR-Net and HDRStats-Net), and the incorporation of the *HDR*s into university curricula worldwide.

Methodology and structure

The study employs a number of qualitative and quantitative methods for this examination. Alongside a review of 24 global *HDR*s since 1990 (including their major concepts, measurement tools, and policy recommendations) and discussions with international policy-makers, scholars, and policy analysts, relevant international policy documents were collected and analyzed, including from the United Nations, the World Bank, the International Monetary Fund (IMF), the World Trade Organization (WTO), the Organisation for Economic Co-operation and Development's Development Assistance Committee (OECD-DAC), and the G20/G8. In addition, to track human development trends (including data used for human development composite indices) and to assess the overall impact of human development policy priorities, in both shaping an international political consensus and in terms of actual implementation, we drew extensively on secondary literature. The book's subsequent chapters are structured as follows.

Chapter 1 considers the conceptual antecedents to human development, such as redistribution with growth, the basic needs approach, and the capability approach, and how these concepts often originated from and shaped global institutions. It further provides a snapshot of how human development evolved within and informed global governance in the 1990s and early 2000s. As a scene-setter for the chapters to follow, Chapter 1 outlines the book's chief research questions, as well as the considerable challenges faced today by the human development paradigm within global governance.

Chapter 2 examines specific *Human Development Report* reform proposals that sought to influence the UN world conferences of the 1990s, including the "20/20" proposal (allocating 20 percent of national and aid budgets for human development priorities), the mobilization of a peace dividend, and the creation of a UN Economic Security Council (ESC). Evidence is shared on how global—and subsequent regional, national, and sub-national—*HDR*s during this period shaped how development is understood, contributing to a paradigm shift in global governance. Their impact on the UN's institutional culture and programming priorities is also considered. In this regard, criticisms of the UN Development Programme (UNDP)—especially in the early days of the *HDR*—to allow its staff the autonomy to criticize governments, and the lessons this holds for other parts of the UN system, are highlighted. The chapter further considers the application of human development policy prescriptions in international policy dialogues beyond the UN proper, including with respect to international financial institutions, the WTO, OECD-DAC, and G8/G20. Special reference is made to the policy and other challenges confronted in pursuing human development reforms and conceptual innovations in the early years of *HDR*s.

Chapter 3 highlights how human development proponents designed evaluation mechanisms—in the form of composite indicators—to quantify social and economic progress and to compare social groups within and between countries over time. Specifically, they sought to respond to the constraints of Gross National Product (GNP) as a measure of human progress, which was viewed as one-dimensional, failing to recognize non-monetized activities, and reflecting market prices in solely monetary terms—thereby remaining silent about the distribution, character, or quality of economic growth. An alternative measure, from the very first global *HDR*, was the Human Development Index (HDI), which began as a simple aggregate of indicators reflecting three major components of human development: longevity, knowledge, and command over resources needed for a decent living or income. The strengths and weaknesses of the HDI as a measure of human progress in global policy debates are examined, as well as the value of related composite indices such as the Gender-related Development Index (GDI), the Gender Empowerment Measure (GEM), the Political Freedom Index (PFI), and the Human Poverty Index (HPI) followed later by the Multidimensional Poverty Index (MPI). In particular, the chapter considers the extent to which human development measurement tools are translated, for example, into revised national budgetary allocations in greater favor of human development policy priorities. The chapter concludes with a discussion on the future of human development composite measures in global governance, including their relationship to the Post-2015 Development Agenda and Sustainable Development Goals.

Chapter 4 applies the lens of critical aspects of human development to explore its evolution in the terms of debates on trade, water, energy, and the environment. It first assesses the relationship between sustainable development and human development, with particular reference to equity, responsibility, consumption, inter- and intragenerational sustainability, and the question of the ends and means of human development. This conceptual relationship has only grown in significance because of increasing concerns about the environment and introduction of the Post-2015 Development Agenda and Sustainable Development Goals. The chapter then offers empirical evidence across a range of global institutions and initiatives, including across several UN agencies and the World Trade Organization, of how the human development school of thought—particularly through the introduction of new norms, measurement tools, and policy prescriptions—affected the international policy discourse on issues relating to access and opportunity (to medicines, water and sanitation, energy, and the global carbon space), rights of the poor and vulnerable communities (of indigenous people, or use of transboundary waters), and enabling conditions for sustainable development (such as capacity for trade, technology, and finance for clean energy, or to build resilience).

In a manner similar to the previous chapter, Chapter 5 also examines the study's primary questions by assessing human development's contribution to: a paradigmatic intellectual shift, an influential international policy agenda-setting role, and innovative reforms to the changing nature of democracy, human rights, and peacebuilding in global and national governance. In doing so, it seeks to demonstrate how human development has benefited from the analytical tools of political scientists, lawyers, and philosophers, thereby enriching the concept of human development and its policy relevance in global institutions. Specific consideration is given to a critical review of the adoption and advancement of policy and institutional reform recommendations from *HDR 2000: Human Rights and Human Development, HDR 2002: Deepening Democracy in a Fragmented World, HDR 2004: Cultural Liberty in Today's Diverse World*, and *HDR 2005: International Cooperation at a Crossroads—Aid, Trade, and Security in an Unequal World.*

To conclude, Chapter 6 proposes a rethink of the ends and means of global governance, identifies four new transitions under way in global governance, and considers several "new frontier" issues for which human development analytical tools and policy guidance can continue to make a contribution in the work of global institutions. Specifically, the chapter offers recommendations on how human development can be further enriched and embedded in international decision-making, to better respond to new risks associated with the international financial crisis, terrorism and other forms of violent extremism (freedom from fear), resource scarcity (freedom from want), and leadership and structural concerns within the United Nations system. Concrete recommendations for strengthening human development-related concepts, metrics, and policy prescriptions are introduced to renew and extend international political commitments to human development priorities and to avert potential backsliding away from progress made in recent decades.

Notes

1 The points on whether there is a global consensus on development, how it should be done, and whether there is agreement today that it should be done are inspired from an email exchange with the scholar Craig Murphy on 21 April 2015.

2 For example, Sakiko Fukuda-Parr and A.K. Shiva Kumar, eds, *Readings in Human Development: Concepts, Measures and Policies for a Development Paradigm* (New Delhi: Oxford University Press, 2003); Mahbub ul Haq, *Reflections on Human Development* (New York: Oxford University Press, 1995); and Khadija Haq and Richard Ponzio, eds, *Pioneering the Human Development Revolution: An Intellectual Biography of Mahbub ul Haq* (New Delhi: Oxford University Press, 2008).

1 Human development

An idea whose time has come?

- **Conceptual antecedents to human development and their origins in global institutions**
- **Snapshot of human development in global governance over the past two and a half decades**
- **Chief research questions**
- **Why human development faces considerable challenges today in global governance**
- **Conclusion**

The purpose of this book is to examine the idea of human development and its influence, over the past two and a half decades, in redefining our understanding of human progress in institutions of global governance. Beginning with the first *Human Development Report* (*HDR*) in 1990, human development was defined as "a process of enlarging people's choices" to improve the human condition. Over time, it also came to be accepted as an expansion of human capabilities, an enhancement of freedoms, and a fulfillment of human rights. Whereas economic growth schools focus exclusively on only one choice—income—human development embraces the enlargement of all human choices: whether economic, social, cultural, or political. Many scholars, international policy-makers, and development practitioners view human development as both an intellectual and policy breakthrough given its success in reminding us of the ultimate purpose of development: to treat all people—present and future generations—as ends.[1] When healthy, educated, well-nourished, and empowered, people are also the chief means of development.

Thinking on human development has contributed to a shift in academic and policy arenas away from national-income accounting to people, their well-being, and the human capabilities to expand their well-being. This project will assess the influence of the idea of human development in contributing to a paradigmatic shift in global institutions, by analyzing the extent to which international public policy makers now think differently about development, raise new questions, and give priority to human development concerns and the multitude of constraints faced in pursuing this agenda. In particular, it will give attention to the impact of the human development approach in shaping how progress is perceived and measured in global governance in specific issue areas such as democratic governance, energy, the environment, human rights, peacebuilding, and trade. The book will highlight the major human development-related conceptual, measurement, and policy and institutional reform innovations introduced through, in particular, hundreds of noted global, regional, national, and subnational *HDR*s. It will also offer a critical perspective on the limitations to current human development approaches and the obstacles that must be overcome if human development is to remain relevant, let alone gain wider traction, in international policy circles. The human development "brand" in the 1990s was itself a critical response and alternative to the 1980s structural adjustment policies of the international financial institutions, and to a preoccupation by wealthy donor nations with GNP, per-capita income, and other national-income accounting tools.

A combination of the 2008–2009 international financial crisis, terrorism, and other forms of extremism, resource scarcity and climate change, and leadership and structural concerns within the chief global institutional champion of human development—the United Nations (UN)—have converged today to place at risk both a continued commitment to human development priorities and the tangible gains achieved in recent decades. In short, a reinvigoration of international policy debates about human development is vital to rejuvenate and sustain global institutions, such as the UN, World Bank, and World Trade Organization (WTO). This is especially the case with regards to "new frontier issues," such as energy, the environment, and democratization at all levels of governance.

This chapter defines and presents the key features of human development, beginning with a review of the conceptual antecedents to human development and their origins in global institutions. It then provides a snapshot of human development policy and institutional reform priorities in global governance over the past two and a half decades, giving attention to what has changed and what has not changed in global human development policy debates. The chapter then elaborates on four fundamental questions the book intends to address. First is the issue of paradigm shift: has human development transformed how development is understood in global governance? Second is the issue of agenda-setting: to what extent have international policy-makers sought to advance human development reform priorities in international policy forums? Third entails new horizion issues: what are the emerging human development ideas and innovations in global institutions? And the fourth concerns obstancles: what constraints and challenges—whether policy, resource, technical, or others—were confronted or are expected in pursuing this agenda?

While citing the phenomenal growth in human development-related policy reports, coupled with ample references to the concept since the UN world conferences of the 1990s, initial evidence cited in this chapter suggests that a number of factors preclude human development from deepening its roots in global institutions and international decision-making more broadly. Indeed, the concept is under immense strain and its associated policy prescriptions face the prospect of serious setbacks if an appropriate set of responses is not fashioned soon.

Conceptual antecedents to human development and their origins in global institutions

The human development conceptual framework can be traced back to philosophical traditions underscoring the sanctity of human life, as espoused by the world's great religions—both from the East and the West—and the ancient writings of the renowned Greek philosophers Plato and Aristotle, and the Indian philosopher Kautilya. In more recent times, human development derives inspiration from classical economics beginning with Sir William Petty's *Political Arithmetic* and Adam Smith's *The Wealth of Nations*, which acknowledged the central role of the ingenuity and hard work of people to a nation's progress.[2]

By the 1950s and 1960s, alongside the meteoric growth in the number and reach of international organizations, attention shifted in influential bodies such as the World Bank and International Monetary Fund to an emphasis on the capital inputs and related macroeconomic reforms and balances to achieve economic growth. Perceived as a "growth only agenda" that appeared to devalue the significance of individual labor inputs and social outcomes, successive challenges were lodged in the 1970s against this perspective by leading scholars and policy analysts. In conjunction with a series of studies commissioned by the International Labour Organization, for example, Dudley Seers critiqued the emerging economic orthodoxy of the international financial institutions by stressing the employment objective, whereas Hollis Chenery, Hans Singer, and Richard Jolly went further in making the case for redistribution with growth as essential to fostering socio-economic development over the medium and long term.

These early intellectual forerunners to thinking on human development were soon followed by the basic needs approach, spearheaded by Paul Streeten, Frances Stewart, and Mahbub ul Haq while at the World Bank in the 1970s, which posits that the poor need certain basic goods and services (e.g., food, water, clothing, and shelter) and a decent life depends on how these are consumed. The basic needs approach further transformed how national and international policy-makers conceived of and measured absolute poverty, by seeking to define the absolute minimum resources (consumption goods mainly) needed for long-term physical well-being. Lists of basic needs in subsequent decades would also include health services, education, and sanitation.

Finally, in the 1980s, Amartya Sen—a long time collaborator with the United Nations—innovated his "capability approach," which relates the evaluation of the quality of life to the assessment of the capability of human beings to function in society. According to Sen, "If life is seen as a set of 'doings and beings' that are valuable, the exercise of assessing the quality of life takes the form of evaluating these functionings and the capability to function."[3] The notion that a person's capability reflects his or her freedom to choose between different ways of living would later form a cornerstone of the human development paradigm.

Building on the above approaches toward fighting poverty, expanding human freedoms, and achieving more balanced, equitable growth, the human development school emerged in the mid-1980s and early 1990s. Alongside its chief architects, Mahbub ul Haq and Amartya Sen, its earliest key proponents included Meghnad Desai, Gustav Ranis, Frances Stewart, Paul Streeten, Khadija Haq, and Richard Jolly. Their vigorous debates in annual meetings of the Society for International Development's "North-South Roundtable" contributed to the initial contours and distinguishing features of what would later encompass a robust, yet somewhat elastic human development conceptual framework. For instance, while advocating the need for developing country governments to divest from productive sectors of the economy where the business community maintains clear comparative advantages, they also took collective aim at the shortsightedness and destabilizing effects of the strict structural adjustment programs of the 1980s—introduced by the international financial institutions, with the backing of rich donor nations. At the same time, this eminent group of economists and development specialists stressed the urgency of donors and developing countries alike redoubling public investments in what they deemed as essential "human development priorities." These include basic education (including a working literacy level), primary health care, potable water, basic sanitation, and the protection of the rights of women and children.

From its infancy, the human development paradigm sought to shift how national policy-makers, citizens, and international bodies judge national progress. In incorporating Sen's "capability approach," it also scrutinized the ways and means of enlarging people's choices to live productive and meaningful lives. By the mid-1990s, Mahbub ul Haq observed broad agreement on key aspects of the human development paradigm:

- Development must put people at the center of its concerns.
- The purpose of development is to enlarge all human choices, not just income.
- The human development paradigm is concerned both with building up human capabilities (through investment in people) and with using those human capabilities fully (through an enabling framework for growth and employment).
- Human development has four essential pillars: equality, sustainability, productivity, and empowerment. It regards economic growth as essential but emphasizes the need to pay attention to its quality and distribution, analyzes at length its link with human lives, and questions its long-term sustainability.
- The human development paradigm defines the ends of development and analyzes sensible options for achieving them.[4]

Over time—and as illustrated in this volume—human development has evolved into a rich, multidimensional concept and framework for analysis and understanding development. For example, leading a long and healthy life, being educated, and enjoying a decent standard of living were initially portrayed as the most critical development choices. But, in subsequent years, political freedom, broader human rights, and environmental sustainability grew increasingly important, as detailed in Chapters 5 and 6. This sequencing of development priorities arose, in part, because of criticisms by governments in the Global North and South that objected strongly to initial attempts by the *HDR*s to provide a disinterested, neutral assessment of freedom in countries.

Critical perspectives about the human development conceptual framework and the norms and concrete policy prescriptions that flow from it were introduced in the early years by scholars and policy analysts, such as Leen Boer, Allen Kelly, Ad Koekkoek, and V.V. Bhanoji Rao.[5] In particular, they criticized the elasticity of the basic concept of human development—for example, that it tries to "mean everything to everyone" and, therefore, lacks analytical utility or the ability to facilitate policy consistency—as well as the over-simplification of complex policy problems because of the framework's lack of methodological rigor and reliable data sources.

However, with each successive global, regional, national, and subnational *HDR*, combined with writings associated with the *Journal of Human Development and Capabilities* and the global membership of the Human Development and Capability Association, the human development conceptual framework and its associated analytical tools and policy prescriptions were further refined through stringent "field testing" in diverse policy settings. In this volume, we examine the application and impact of human development in policy deliberations undertaken in institutions of global governance. Although the school of thought faces innumerable intellectual and political challenges today, human development has demonstrated consistently, during the past two and a half decades, a marked and potentially durable effect on international policy discourse related to fundamental questions of world order, organization and social spending within states, and the balancing of competing political, economic, social, and cultural priorities.

Snapshot of human development in global governance over the past two and a half decades

Besides the annual global *HDR*, more than 700 national and regional *HDR*s have been produced since 1990, and a large literature has emerged on the finer aspects of human development theory and measurement. Human development—as advanced through the global *HDR* and other vehicles—has shaped international policy discourse over the past 25 years in institutions of global governance in at least three concrete ways. First, human development has shifted and expanded what is commonly perceived as the goal of development today—a new conceptual framework or paradigm of human development. Second, it has innovated new means for gauging human and national progress (new human development measurement tools). Third, it has introduced many recommendations for change in international public policy (new global governance policy and institutional reform proposals). A few examples follow of human development's impact, detailed at length in this volume.

Shifting and expanding what is commonly perceived as the goal of development today

While emanating from the traditional development community, the *HDR*s had a profound effect on thinking and practice related to the conflict-security-justice nexus in global governance shortly after their birth. The concept of human security, introduced in the 1993 and subsequently elaborated in the *1994 Human Development Report*, soon shaped how policy-makers view the origins of, and an appropriate international response to, deadly armed violence.

The *HDR*s defined human security as "people exercising their human development choices safely and freely" and "ensuring freedom from fear and freedom from want." In this way they contributed to a broader understanding of the sources of violent conflict and the need to inject targeted resources for human development into any kind of strategic framework or individual (sectoral) measures aiming to cope with violent conflict and reduce the likelihood of its recurrence. Both human development and human security placed people at the center of future conceptions of peacebuilding, serving as both the ends and chief means towards achieving sustainable peace. Consequently, they markedly changed the international mindset on the requirements for effective peacebuilding. This mindset was freed in part by the lapse in the superpower (and modern military and state-centric) standoff that virtually paralyzed global institutions such as the UN during the Cold War.

At the 2005 UN summit, world leaders stressed in a section on "human security" in the outcome document "the right of all people to live in freedom and dignity, free from poverty and despair." They also recognized that "all individuals, in particular vulnerable people, are entitled to freedom from fear and freedom from want, with an equal opportunity to enjoy all their rights and fully develop their human potential."[6] World leaders also endorsed for the first time the norm of the "Responsibility to Protect" populations from genocide, war crimes, crimes against humanity, and ethnic cleansing.[7]

Innovating new means for gauging human and national progress

In terms of measurement tool innovations, the rich concept of human development and its associated policy reports shifted the development discourse in global institutions away from a narrow economic criterion as the chief way to measure a nation's progress to include health and knowledge indicators. This was achieved, in particular, through the Human Development Index (HDI), which measures the average achievements in a country in three basic dimensions of human development: a long and healthy life, access to knowledge, and a decent standard of living. To enable cross-country comparisons, the HDI is, to the extent possible, calculated based on data from leading international data agencies and other credible data sources available at the time of writing. Over time, new measurement tools for measuring gender (1995) and the multiple dimensions of poverty (1997) were introduced, and the political dimensions of human development were also fully elaborated in path-breaking global *HDR*s on human rights (2000) and democratization (2002).

A key strength of the HDI is that it seeks to measure progress by countries—but also sub-national, regional, and global units of governance—in advancing the comprehensive, holistic, and integrative concept of human development. While Mahbub ul Haq, Amartya Sen, and other authors acknowledged that the HDI is a crude instrument that does not represent the totality of what constitutes human development, they felt, nevertheless, that it was a vast improvement over the even more crude national-income accounting tools of GNP and Gross Domestic Product (GDP). The vast popularity of the HDI since its inception—as an integral part of the UN system's most widely cited annual report—and its application in disaggregating data at various sub-national governance levels meant that human development trends and change over time could be presented with relative ease in multiple ways.

Introducing many recommendations for change in international public policy

The subsequent chapters provide an overview of human development policy and institutional reform priorities in global governance over the past two and a half decades, giving attention to what has changed and what has not changed in global human development policy debates. Beyond influencing policy discussions at UN headquarters in New York and Geneva, we argue that the *HDR*s made headway in articulating a coherent response and intellectual counterweight to the highly influential "Washington Consensus" set of policy priorities advanced by global institutions such as the World Bank and IMF. Unlike the neo-liberal approach adopted by Washington Consensus advocates, which placed confidence in the trickling-down of wealth in society, a human development approach assumes that economic growth must be made consciously pro-poor and pro-people and, therefore, a "goal-oriented" rather than "growth-oriented" poverty strategy is needed.

Besides senior World Bank officials such as James Wolfensohn beginning to cite human development regularly in the late 1990s, during the 2000s a new "network" for human development was established as the World Bank's home for policy, programs, and research in the fields of education, health, social protection, and labor. Albeit adopting a significantly narrower definition of human development than UNDP and *HDR* authors, the creation of, for example, senior positions dedicated to human development has contributed to increasing concern for poverty reduction and strengthening human capabilities at the World Bank. It is also no coincidence that the World Bank shifted away from a core focus on top-down structural adjustment programs to more participatory *Poverty Reduction Strategy Papers* (*PRSP*s), between the 1980s and the early 2000s, as interest in the concept of human development moved to the fore internationally. Unlike structural adjustment, *PRSP*s place a premium on people's participation and the centrality of human agency for tackling the causes of human poverty, common themes in all *HDR*s, although perhaps amplified in the 1993 and 1997 annual reports.

In addition to this broad economic and development policy debate, successive *HDR*s have advocated for the following three global institutional reforms, which have garnered traction in international policy circles. The first is increasing civil society organization participation in the discussions of the UN General Assembly, Economic and Social Council (ECOSOC), and Security Council. Except perhaps for the Security Council this has become a growing reality.[8] The second is expanding the global economic decision-making clout of developing countries by transforming ECOSOC into a powerful 22-member Economic Security Council. This idea, one can argue, has been taken on board by the G20 and its heads of government level meetings since 2008. Third is that the *HDR*s have also focused global attention on making the World Bank and IMF more accountable for their actions to board members and to the people affected by their actions. They have also focused on improving WTO consultations, discussions, negotiations, and decision-making so as to make the organization more transparent, participatory, and democratic.

Chief research questions

The book addresses four fundamental questions. First, has human development transformed how development is understood in global governance? The book explores the issue of whether human development has contributed to a paradigmatic shift in development thinking within global institutions, from an emphasis on GDP growth and per capita income as the chief barometers of a nation's progress to a broader notion that balances economic with social, cultural, and political dimensions of development. In doing so, human development attempts to shepherd a return to a more holistic understanding of human well-being and agency as advocated by classical economists of the late eighteenth and early nineteenth centuries. To aid this effort, consideration will be given to the application within global institutions of new composite indices for measuring and better understanding the breadth and scope of human development.

Second, to what extent have international policy-makers sought to advance human development reform priorities in international policy forums? Citing empirical evidence from international conferences, periodic meetings of global bodies, international trade negotiations, and other international policy forums over the past two decades, the study considers whether and to what extent a human development policy framework, measurement tools, and specific policy priorities (e.g., redirecting investments in national armaments toward investments in a nation's citizens) shaped the international policy agenda of the 1990s and early 2000s.

Third, what are the emerging human development ideas and innovations in global institutions? With an eye towards both the immediate and medium-term future, the book explores possible "new horizon issues" where the adoption of a human development approach in global institutions can offer different and unique ways to tackle a particular global governance problem-set. By factoring in inequality concerns and the need to ensure a political voice for citizens at all levels of governance, for example, human development can continue to pioneer innovative solutions to intractable international issues, including the global financial crisis, violent political and religious extremism, and environmental decay.

Fourth, what constraints and challenges—whether policy, resource, technical, and others—were confronted or are expected in pursuing this agenda? This study analyzes the various obstacles to adopting and then implementing human development reform priorities, including alternative (competitor) policy frameworks. Within a human development policy framework as played out in global institutional debates, the book considers the factors for overcoming constraints and challenges to human development in some areas, while progress in others remains inhibited.

Why human development faces considerable challenges today in global governance

Despite phenomenal growth in the number and influence of global, regional, national, and sub-national *HDR*s, human development—as an intellectual and policy movement—has come under significant strain in recent years in institutions of global governance. The 2008–2009 financial collapse in the United States rippled through the global financial system, affecting most regional and national economies adversely—especially highly exposed and already debt-ridden European economies. The steady economic downturn, the steepest fall recorded in global markets since the 1930s, has placed at risk both the political consensus and ability of advanced industrialized countries to guarantee even the most basic investments in their citizens. Meanwhile, as emerging economies and less developed countries strive, in part, to achieve the quality of life standards of the industrialized world, they are receiving mixed signals about whether costly investments in human development are sustainable and can yield positive economic and other dividends.

On one hand, as Chapter 6 notes, considerable progress has been achieved toward the realization of most of the eight Millennium Development Goals (MDGs) for 2015, including the (arguably most) critical MDG Goal No. 1 of ending poverty and hunger. On the other, many analysts attribute the marked success to, in particular, the phenomenal growth rates over the past decade of the world's two largest countries, China and India. Meanwhile, more than a billion citizens associated with fragile or conflict-affected countries continue to struggle to meet their basic human needs,[9] let alone meet the more ambitious targets associated with the 15-year MDGs project. With a few notable exceptions where large aid programs are sustained for years because of a recipient country's links to security concerns found within Western donor countries, many vulnerable countries struggle to cope when the cameras turn to a new "hot spot," international peacekeeping is drawn down, and donor assistance dries up.

Many—if not a great majority—of developing countries are committed to shifting to an emphasis on "trade not aid." Indeed, over the past two decades, they have become champions of the notion of a "new framework for development cooperation" (see Chapter 2) as advocated in the earliest *HDR*s, calling for North-South exchanges in ideas, people, technology, goods, services, and foreign direct investment, rather than a preoccupation with foreign aid. But, as detailed in Chapter 4, the on-going WTO Doha Development Round continues to come up short in terms of liberalizing trade in two key sectors where many developing countries maintain a distinct competitive advantage: agriculture and textiles.[10] And despite noble intentions, the UN Peacebuilding Commission, established in 2005, has demonstrated limited empirical evidence that countries placed before it—often in the aftermath of a UN peacekeeping mission drawdown—have benefited in tangible and sustained ways which expand human development opportunities in fragile or conflict-affected states.[11]

As explained throughout the book, today's new and emerging international issues threaten to take human development off the global agenda, undermining the global institutions most closely associated with the human development school's progressive international reform agenda. The awesome effects of rising sea levels because of global climate change are expected to overwhelm coastal communities, as well as decrease access to potable water for millions of people. Rapid rural to urban migration continues unabated as many of the planet's seven billion (and growing) inhabitants seek improved living conditions. Yet, more often than not, this migration further stresses over-stretched urban infrastructure and services.

In addition, defensive responses to continued threats posed by militant extremist groups in rich and poor nations alike continue to divert hundreds of billions of dollars from productive economic and social development endeavors that could lift many from abject poverty. Every day, the competence and legitimacy of global institutions with a mandate for responding to these challenges is placed under significant strain, particularly in the eyes of citizens from growing countries experiencing economic change and new political freedoms.

Partly in response to the underperformance of current institutions of global governance to secure and sustain livelihoods for the vast majority of the planet's citizens living in the developing countries of the Global South, ideas for new global policy-making forums and implementing bodies continue to gain attention, especially when an idea's proponents are among the most powerful emerging economies in the world. For example, at the fourth annual summit of the BRICS (Brazil, Russia, India, China, and South Africa), held in March 2012 in New Delhi, the national leaders assembled called for the establishment of a new development bank.[12] Put forward initially by India, the proposed "BRICS-led South-South Development Bank" has three primary aims. First, to provide capital for development, including infrastructure, projects in developing countries. Second, to lend, in the long term, during global financial crises (including to Western countries). Third, to issue convertible debt, which could be bought by the central banks of member countries to reduce risk. At the same time, the efficacy and resilience of today's global financial architecture to prevent or withstand, let alone cushion against, future economic shocks continues to be called into question. Causes here are the continuation of the economic crisis in the Eurozone countries and the fragile economic recovery under way in North America, the United Kingdom, and Japan.

Conclusion

Despite fundamental advances in human development globally over the past six decades—including an acceleration of progress in many developing countries during the last two decades—heightened human insecurities foment democratic disillusionment and raise new questions about the legitimacy and effectiveness of global institutions to advance human development policy priorities. Such insecurities are fueled by economic uncertainty, continued violence from international terrorism and civil wars, environmental and energy concerns, and perceived signs of growing inequality in both industrialized and developing countries.

However, as argued in Chapter 6, the wrong lessons are still being derived from the 2008–2009 global financial and subsequent economic crisis and other perennial challenges. Rather than losing faith in more inclusive institutions, a deepening of democracy at all levels of governance—harkening back to the chief recommendations of *HDR 1993* and *HDR 2002*—is needed urgently, coupled with larger and better targeted investments in human development.

Especially in democratic political systems, policy-makers need to demonstrate that they can deliver public goods for the betterment of society. From the early 1990s, human development offered global institutions—during a period of renewal and great promise—a comprehensive development agenda to help policy-makers improve the quality of their citizen's lives, thereby enhancing their own legitimacy and respect as leaders. In particular, human development made addressing the neglected or "missing" dimensions of development far more convincing by making the case that it is in the short- and long-term interest of the well-off, within the North and South, to tackle these issues head-on as the world integrates at breakneck speed.

This book underscores the sobering challenges to operationalizing human development through both global and national institutions today. It also points, however, to key elements of how human development was presented in the 1990s and early 2000s to guide lessons for taking the human development paradigm forward in the present global context. Such elements include leadership, agenda-setting, evidence-based policy, timing, coalition-building, and campaigning.

This volume explores lessons learned from past successes while seizing new opportunities created by, for example, new technologies and the social media. It aspires to innovate the business model required to bring back to the international agenda the "missing dimensions" of development that sat squarely at the center of UN world conferences and other global policy forums in the early- to mid-1990s. In revisiting past successes and core characteristics of the human development school of thought, a case is made for why human development remains relevant in an era of accelerated globalization. However, while citing the phenomenal growth in human development-related policy reports, coupled with the concept's ample references since the UN world conferences of the 1990s, evidence suggests that a number of factors preclude human development, in its present formulation, from deepening its roots in global institutions and international decision-making more broadly. Indeed, the rich and forward-looking concept and its associated policy prescriptions face the prospect of further setbacks if not re-evaluated and refashioned soon.

Notes

1 *HDR 2002* further elaborates on this definitional point by stating: "Human development is about people, about expanding their choices to lead lives they value. Economic growth, increased international trade and investment, technological advance – all are very important. But they are means, not ends. Whether they contribute to human development in the 21st century will depend on whether they expand people's choices, whether they help create an environment for people to develop their full potential and lead productive, creative lives." UNDP, *Human Development Report 2002: Deepening Democracy in a Fragmented World* (Oxford: Oxford University Press, 2002), 13.

2 As Amartya Sen contends, "At the technical level, William Petty did, in fact, pioneer the measurement of the gross national product, the GNP, using both 'the income method' and 'the expenditure method' of estimating national income. This was a very significant contribution to applied economics, and yet at the same time, Petty was quite clear that the interest in incomes and expenditures lies not in themselves but in their serving as important means to more profound ends [such as 'the Common Safety' and 'each Man's particular Happiness']." Amartya Sen, "Foreword," in *Readings in Human Development*, ed. Sakiko Fukuda-Parr and A.K. Shiva Kumar (New Delhi: Oxford University Press, 2003), viii. Separately on Adam Smith, Sen writes, "The development of human capability in leading a worthwhile life as well as in being more productive is quite central to Smith's analysis of 'the wealth of nations'." Amartya Sen, "Human Capital and Human Capability," *World Development* 25, no. 12 (1997): 1959.

3 Amartya Sen, "Development as Capability Expansion," *Journal of Development Planning* 19 (1989): 43.

4 Mahbub ul Haq, *Reflections on Human Development* (New York: Oxford University Press, 1995), 21.

5 V.V. Bhanoji Rao, "Human Development Report 1990: Review and Assessment," *World Development* 19, no. 10 (1991): 1451–60; Leen Boer and Ad Koekkoek, "Human Development Report: Fad or Fixture?" *Development Policy Review* 11, no. 4 (1993): 427–38; and Allen Kelly, "The

Human Development Index: 'Handle with Care,'" *Population and Development Review* 17, no. 2 (1991): 315–24.

6 United Nations, "2005 World Summit Outcome," UN doc. A/RES/60/1, 24 October 2005, para. 143, www.un.org/womenwatch/ods/A-RES-60-1-E.pdf.

7 Ibid., paras. 138, 139.

8 The United Nations has increased efforts to include civil society in its work. In the 1990s, the UN World Conferences enhanced the voices of civil society within the UN system. The General Assembly, ECOSOC, the Peacebuilding Commission, and the UN Global Compact have also reached out to broader constituencies by making an effort to formally consult and engage with civil society organizations and the private sector. This outreach towards a wider segment of the world population (and far beyond government representatives), although still far from perfect, can be seen as some limited progress towards the democratization of the world organization's work, as strongly advocated by multiple *HDR*s.

9 This thesis is fully developed in, for example, Paul Collier, *The Bottom Billion: Why the Poorest Countries are Failing and What Can be Done About It* (Oxford: Oxford University Press, 2007).

10 In fact, as argued in *HDR 1992* and subsequent reports, restricted global markets in key agriculture and textile sectors, coupled with barriers to immigration, costs developing countries $500 billion a year—far more than they receive in foreign aid. UNDP, *Human Development Report 1992* (New York: Oxford University Press, 1992), 5.

11 For example, see Richard Ponzio, "After Exit: The UN Peacebuilding Architecture," in *Exit Strategies and State Building*, ed. Richard Caplan (New York: Oxford University Press, 2012). At the same time, if the objective of the UN Peacebuilding Commission (PBC) is mainly to prevent the recurrence of violent conflict (which, in fact, only represents a portion of the new intergovernmental advisory body's mandate, as outlined in the PBC's General Assembly and Security Council founding resolutions of 2005), it has fared reasonably well. This despite a brief outbreak in violence in Burundi, in April 2008 and again in 2015, and a military coup d'état in Guinea-Bissau, in December 2008 and April 2012.

12 Interestingly, the annual BRICS Summit and discussion on the idea of a new development bank coincided with the selection of a new World Bank president. For this position, as has become custom, the US candidate (Dr. Jim Yong Kim) was endorsed by the bank's Board of Executive Directors—where Western countries continue to exert significant influence in decision-making. However, in a manner unprecedented for the selection of World Bank presidents, a new level of competition was introduced into the process when the finance minister of Nigeria at that time (Dr. Ngozi Okonjo-Iweala) and the former finance minister of Columbia (Dr. José Antonio Ocampo) were also interviewed by the board.

2 The international policy impact of the human development approach in the 1990s … the early years

- **Crashing the global stage: the**
- **Rationalizing resources through the 20:20 Compact**
- **Mobilizing a global peace dividend**
- **Establishing an Economic Security Council**
- **Changing institutional culture: human development and UN system personnel**
- **Challenging the Washington Consensus: human development and international policy dialogues beyond the UN proper**
- **A "New Framework for Development Cooperation"**
- **Conclusion**

The human development approach or school of thought emerged, in part, in response to the structural adjustment programs of the 1980s to mitigate the debt crisis afflicting much of the developing world. In particular, scholars and practitioners associated with the North-South Roundtable introduced, in the mid-1980s, the concept of human development in a series of roundtable reports that both analyzed and critiqued the austerity measures advocated by the International Monetary Fund and donor countries. More significantly, human development advocates offered a positive, alternative vision—among international and national policy-makers, as well as global civil society—for what development means and should emphasize in policy and programmatic terms near the turn of a new century. Starting in 1990, the *Human Development Report* (*HDR*), which soon became the United Nations Development Programme (UNDP)'s flagship annual publication, served as the chief platform for advancing this vision. And the first major test, in terms of whether the analysis and recommendations of the *HDR*s could gain international policy traction, played out in connection with the United Nations (UN) global development agenda in the immediate aftermath of the Cold War.

This chapter examines specific *HDR* reform proposals that sought to influence the UN world conferences of the 1990s, including the 20:20 proposal (allocating 20 percent of national and aid budgets to human development priorities), the mobilization of a peace dividend, and the creation of a UN Economic Security Council. Evidence is shared on how global—and subsequent regional, national, and sub-national—*HDR*s during this period shaped how development is understood, contributing to a paradigm shift in global governance. Their impact on the UN's institutional culture and programming priorities is also considered. In this regard, criticisms of UNDP—especially in the early days of the *HDR*—to allow its staff the autonomy to criticize governments, and the lessons this holds for other parts of the UN system, are highlighted. The chapter further considers the application of human development policy prescriptions in international policy dialogues beyond the United Nations, including within the international financial institutions and forums: the World Trade Organization, the OECD-DAC, the Group of Eight major industrialized countries (G8), and the Group of Twenty (G20). Special reference is made to the policy and other challenges confronted in pursuing the human development agenda in its early years.

Crashing the global stage: the *HDR*s and the UN world conferences of the 1990s

When the global *HDR* series was launched on 24 May 1990 in London, the time was ripe to reconsider the fundamental purpose and policy priorities associated with development assistance from the industrialized countries of the "Global North" to the developing countries of the "Global South." The Berlin Wall had crumbled the previous year under the weight of millions of citizens living behind the Iron Curtain with a yearning for political and economic freedom. With the sudden dampening of foreign aid's national security imperative, coupled with the growing chorus of voices challenging the efficacy of sharp cutbacks in public social investments during a long period of economic malaise, human development would quickly provide an alternative yet pro-economic-growth approach to development, responding to many of the inherent shortcomings and contradictions accompanying conventional thinking.

The *Human Development Report 1990* defined human development as "a process of enlarging people's choices. The most critical of these wide-ranging choices are to live a long and healthy life, to be educated and to have access to resources needed for a decent standard of living. Additional choices include political freedom, guaranteed human rights and personal self-respect."[1] By the time the concept was presented in the inaugural *HDR*, human development had been deliberated on for several years by Mahbub ul Haq, Khadija Haq, Uner Kirdar, Amartya Sen, Richard Jolly, Frances Stewart, Paul Streeten, and other internationally renowned scholars and practitioners associated with the North-South Roundtable. Arguing that "there was no automatic link between economic growth and human development, and that financial budgets ought not to be balanced by unbalancing human lives," the North-South Roundtable promoted the imperative of putting people at the center of development in three successive gatherings in the mid-1980s on human development.[2] As Cold War thinking dissipated in the early 1990s, the human development approach would soon find a global stage through the UN for amplifying this alternative development vision and associated policy prescriptions.

It was perhaps no coincidence that the *HDR* was initiated alongside an unprecedented series of UN-sponsored world conferences, beginning with the 1990 World Summit for Children and soon followed by high-level conferences on sustainable development (1992), human rights (1993), population (1994), social development (1995), women (1995), and urban development (1996). Indeed, for most of these diplomatic gatherings, the *HDR*s served as an authoritative analytical and advocacy tool to both better understand and respond to acute development gaps.[3] While it is perhaps the annual global Human Development Index ranking and associated composite indicators for which the report is most well-known (see Chapter 3), several innovative reform initiatives shaped deliberations at the UN world conferences of the early 1990s and beyond. Among the most important, based on their success in spurring international policy dialogues, catalyzing pressure groups outside of governments and international organizations, encouraging unorthodox conceptual creativity, and influencing policy change, were the 20:20 Compact, mobilizing a Global Peace Dividend, and establishing an Economic Security Council.[4] They are discussed in turn below.

Rationalizing resources through the 20:20 Compact

Building on earlier *HDR*s (especially *HDR 1991* on financing human development), the "20:20 Global Compact for Basic Social Services" was formally introduced in *HDR 1994*, on human security, and subsequently elaborated in Mahbub ul Haq's book *Reflections on Human Development*.[5] Rather than calling for new funds, the 20:20 Compact advocates that 20 percent of developing country budgets and 20 percent of industrial country aid be allocated to human priority expenditure, such as education, health, family planning services, nutritional programs, and low-cost water supply and sanitation. Based on a notion of shared responsibility between donor and recipient countries, this simple yet compelling proposal reflects the decent human development levels achieved by countries that allocate, on average, 20 percent of public spending on human priorities.[6] It also aims to mitigate against excessive military spending, investments in loss-making public enterprises, and diverting scarce donor resources to wasteful prestige development projects. In sum, if adopted, the proposal was estimated to redirect US$30–40 billion annually, during the period 1995–2005, to tackle the worst forms of human poverty.[7]

In the lead-up to the 1995 World Summit for Social Development, the 20:20 Compact garnered considerable policy attention and references were made to the proposal in both the summit's main outcome document ("Programme of Action") and an unofficial Copenhagen Social Charter. As stated in the summit's Programme of Action, "Agreeing on a mutual commitment between interested developed and developing country partners to allocate, on average, 20 percent of ODA [overseas development assistance] and 20 percent of the national budget, respectively, to basic social programmes."[8] However, given the deliberate use of the term "interested" in the Programme of Action and the non-binding nature of the Copenhagen Social Charter, UN member states participating in the summit had limited incentives or sanctions to alter their public spending and aid practices.

Perhaps a far more significant impact came a few years after the 1995 summit, with the articulation of the UN's Millennium Development Goals and subsequent endorsement by world leaders at the 2000 UN Millennium Summit. The MDGs outline specific, time-bound targets focused on reductions in, for example, human poverty, illiteracy, child mortality, and gender inequality. The 20:20 proposal served, in part, as a forerunner to the highly successful Millennium Development Goals by rallying both rich and poor nations around their mutual interest in combating several closely linked human deprivations simultaneously. Before the proposal's formal introduction in the *HDR 1994*, the annual reports between 1990 and 1993 also called for poverty-focused international development agendas based on a compact between developed and developing countries. These too heralded the Millennium Development Goals that emerged later in the decade.

Although institutional follow-up on the original 20:20 Compact idea may have been limited until its associated commitments were integrated into well-resourced programs associated with the implementation of the MDGs, the proposal gained significant policy traction, in the mid-1990s, because it offered a concrete response to a number of key global targets endorsed by the international community at several UN world conferences from the early to mid-1990s.[9] First and foremost, the 20:20 proposal initiated a shift in thinking about the importance of adopting a unified approach and integrated intellectual framework, among donors and developing countries alike, around a defined set of human development priorities. By the mid- to late-1990s, the intellectual shift informed a policy shift that contributed, in part, to the advent of the Millennium Development Goals for 2015. By focusing donor assistance and public spending in the Global South on a common set of human priorities, the MDGs have largely framed the development policy dialogue during the first 15 years of the new century.

Mobilizing a global peace dividend

With the thawing and then complete transformation of Cold War relations, the *HDR* series championed the need to reorder military budget priorities in both the developing and industrialized worlds. "Military expenditures of the developing countries have increased 7.5 percent a year during the past 25 years, far faster than military spending in the industrial countries," stated the inaugural edition of the report in 1990.[10] By *HDR 1992*, the reports had begun to explicitly proclaim the need for all countries, rich and poor, to commit themselves to peace dividends. It was estimated that military expenditure reductions of at least 3 percent a year in the 1990s could yield by the year 2000 a global peace dividend of around US$1.5 trillion ($1.2 trillion in the industrial countries and $279 billion in developing states).[11] *Human Development Report 1994* prominently featured a recommendation by Nobel Peace Prize Laureate Oscar Arias to create "A Global Demilitarization Fund." Besides linking the reduction in military expenditure to the consolidation of world peace, the proposed fund could aim to reward developing countries that disarm, demobilize, and reintegrate their armed forces back into society; promote arms control; and encourage civic education and democratic participation.[12]

Given that an estimated 25 percent was trimmed from defense budgets worldwide between 1987 and 1994, policy advocates from within the *HDR* community are not in a position to take credit for the significant cuts of around 4 percent per year led by the United States and former Soviet Union.[13] On the other hand, beginning in 1990, the *HDR*s reminded international policy-makers at UN world conferences and other international forums that the savings accrued from military reductions are not automatically converted into human development gains. The former Soviet Bloc countries saw their military spending savings wiped out in coping with their sustained economic crisis, while the United States diverted its savings to drawing down its national debt. At the same time, the *HDR*s sounded the alarm repeatedly about the continued excessively high military spending levels in many of the world's poorest countries, located in South Asia and Sub-Saharan Africa. They also highlighted the combined development and security risks associated with a "global arms bazaar" fueled by cheap surplus weapons in the aftermath of the Cold War.

From the adoption of the 2001 United Nations "Programme of Action to Prevent, Combat, and Eradicate the Illicit Trade in Small Arms and Light Weapons" to the adoption of peacebuilding and statebuilding goals (alongside the socio-economic MDGs) by donors and conflict-affected states through the 2011 "New Deal for Engagement in Fragile States,"[14] many of the themes and recommendations for greater transparency in arms sales and rationalization of military expenditure taken up in successive *HDR*s are slowly coming to fruition. Even elements of Oscar Arias' proposed Global Demilitarization Fund are reflected in the UN Peacebuilding Fund,[15] established in 2006 alongside a new UN Peacebuilding Commission and UN Peacebuilding Support Office. With a growing number of multilateral and bilateral development aid providers offering technical assistance on security sector reform in developing societies,[16] the human development community has helped to insert military spending and the tools of war as a legitimate issue within the wider development dialogue.

And despite startling acts of international terrorism over the past decade, perhaps the most promising trend since the end of the Cold War is the steady decline in both inter- and intrastate political violence.[17] Not coincidentally, this phenomenon has occurred alongside the marked shift in emphasis within global policy dialogues away from a preoccupation with military-led state security to a more balanced conception of security where both the security and welfare of the individual citizen matters. The concept of human security, presented initially in the *Human Development Report 1993*, posits the need to change how security is understood: "from an exclusive stress on national security to a much greater stress on people's security, from security through armaments to security through human development, from territorial security to food, employment and environmental security."[18] Similar to the abovementioned Peacebuilding Fund, a UN Human Security Trust Fund was launched in 1999,[19] and other operational measures were subsequently adopted across the UN system, as described later in this chapter, both to promote human security and to leverage potential peace dividends from within countries recovering from protracted violent conflict.[20]

Establishing an Economic Security Council

The idea of an Economic Security Council stems from the identified need for a more manageable apex organization within the UN framework for effective international policy coordination and leadership on macroeconomic and related matters, including global poverty, food security, labor migration, and environmental safeguards. From its introduction as a "Development Security Council" in *Human Development Report 1992* [21] to its reintroduction in the 1994 report, the Economic Security Council concept responds to the critique—particularly among Western industrialized countries—that the UN Economic and Social Council's 54-member structure is too large and unwieldy. Specifically, the ESC's proposed "membership could consist of 11 permanent members from the main industrial and more populous developing countries. Another 11 members could be added on a rotating basis from various geographical and political constituencies."[22] Besides reviewing threats to international economic stability and agreeing on required action, the Economic Security Council could act as a watchdog vis-à-vis the policies and programs of international and regional financial institutions.[23]

Although the idea of a Development and then Economic Security Council informed vigorous debates on the future of the UN at the time of the 1992 Earth Summit, 1995 Social Summit, and 50th Anniversary of the UN in 1995, this global governance institutional reform has yet to take root in its original, ambitious form in the UN context. However, in lieu of this UN system structural reform, several developments over the past 15 years reflect many of the substantive aims of the ESC. For instance, in January 2000 the UN Security Council convened an unprecedented session on the HIV/AIDS pandemic and its impact on peace and security in Africa.[24] Later that year, Security Council members deliberated on the social, economic, and political impact of violent conflict on women and girls.[25] Responding to performance criticisms, ECOSOC has worked to streamline and focus its agenda over the past decade, as well as to ensure greater sensitivity in its decision-making to the interests of industrialized nations, emerging economic powers, and less developed countries. Finally, the Economic Security Council proposal contributed, over time, to World Bank and International Monetary Fund reform discussions, intensifying pressures to expand representation and voting rights to non-OECD countries based in the developing world. In particular expanding them to the larger BRICS economies: Brazil, Russia, India, China, and South Africa.

While the Economic Security Council concept never took hold in the UN proper in New York, its more or less envisaged membership and substantive focus emerged by 1999, when Canada proposed and the first meeting was held, in December 1999, of the Group of Twenty (G20) Finance Ministers and Central Bank Governors. Collectively, the G20 accounts for around 80 percent of Gross World Product, 80 percent of world trade, and two-thirds of world population. G20 leaders began meeting in November 2008 in Washington, DC at the heads of state and government level, and by the September 2009 gathering in Pittsburgh, the rich countries represented in the G8 announced that the G20 would now serve as the main economic council for large wealthy nations. Admittedly, concerns abound about the G20's ability to enforce its commitments, as well as its overall legitimacy given its informal structure outside of the UN Charter. But the small and manageable size of its influential membership, trust garnered between both its developed and developing members, and timely analysis and recommended responses to urgent economic and related challenges make the G20 a welcome innovation in global governance, consistent with many aspects of the proposed UN Economic Security Council.

Changing institutional culture: human development and UN system personnel

When Mahbub ul Haq initiated the *HDR* in 1990, he did so upon securing agreement on one fundamental condition with the then-UNDP administrator, William Draper III: complete intellectual freedom and editorial independence. Draper, who had been looking outside the UN "to get some intellectual support for what we were doing,"[26] took a significant risk that would later be endorsed by his successors. Within a risk-averse and often conservative UN, such a high degree of independence—for an annual publication that would soon become its most widely read—was unprecedented. Besides representing a paradigmatic shift within the wider development discourse and policy debates, the *HDR*s contributed to a transformation in the institutional culture of the UN system in the 1990s and early 2000s.

Prior to influencing the bureaucratic fabric of the wider UN system, including the international financial institutions, the *HDR*s needed to first secure the trust and respect of the series' primary home institution, the UNDP. This critical relationship, which involved organizational resources, visibility, and prestige, proved beneficial for helping the *HDR*s to capture the immediate attention of policy-makers, scholars, and civil society activists, as well as in shaping international policy debates. But despite the full backing of William Draper and his successor, James Gustave Speth, outreach with other UNDP colleagues in the early years was, at times, bumpy. For one, having the *HDR* drafting team command editorial independence was unique, whether for UNDP or the broader UN system. But it was perhaps Mahbub ul Haq's call for a new way to approach and measure development, combined with his expectation that the UN would naturally champion this new approach, that stirred controversy and dissent within the report's home institution.[27]

Reflecting the skepticism and even outright intellectual hostility toward the earliest reports, Desmond McNeil writes, "the views of UNDP staff and member states were not always positive and supportive of human development. Structural adjustment views were also found in the UNDP in the early 1980s."[28] Plus, some staff questioned how an outsider and a largely non-UNDP group of international advisors could garner remarkable influence within the organization almost overnight.

Some critics were swayed, however, when a UN General Assembly (UNGA) resolution in 1994 endorsed the *HDR*'s editorial independence and applauded the initiation of a new analytical report within the UN system that "was objective rather than ideologically or politically mandated, and that provided valuable information, analysis and policy options."[29] By the mid-1990s, following the UNGA resolution and extensive advocacy efforts undertaken by the *HDR* Office, the idea of human development had been officially accepted within UNDP.[30] According to Craig Murphy, "In the mid-1990s, Draper's successor, Gus Speth, worked with two Secretaries-General and the heads of all the UN programs and funds to make sustainable human development their shared goal, and to establish in-country and global mechanisms for supporting this goal."[31] During a period when the absence of the Cold War had paradoxically threatened foreign aid spending through the UN system and other development assistance providers, the *HDR*s also helped to raise the profile of UNDP vis-à-vis its competition, among both donor and beneficiary countries.

Within a few short years, the concept of human development and its reports at the global, regional, national, and sub-national levels also garnered traction within the wider UN family, including by then-UN secretary-general Kofi Annan (1997–2006), who stated, "[W]e have defined what development means for the individual through our *Human Development Reports* … So we have given a functional and meaningful definition to poverty and development, which wasn't there before."[32] For three astute UN scholars, the comprehensive and integrated perspective offered by the human development framework was its chief comparative advantage for the UN system. In their capstone volume for the UN Intellectual History Project, Richard Jolly, Louis Emmerij, and Thomas Weiss argue, "The most important contribution of the human development approach for the UN as a whole is that it brought together and integrated the four fundamental ideas on which the UN has been founded: peace and negotiation in place of war and conflict; sovereign independence; economic and social development to achieve rising living standards; and human rights for all."[33]

Intellectually—rather than in actual practice—human development represented a breakthrough that allowed senior UN leaders to think, at the very least, in a holistic manner. This could, in turn, begin to break down the conceptual and operational barriers to a more comprehensive approach for tackling the complex and interdisciplinary global governance challenges of the twenty-first century.

There was a push for "One UN" in the early 2000s, including the consolidation of programs, funds, and agencies in a single "UN House" in an aid recipient country and the augmentation of authorities within "special representatives of the secretary-general" (for multidimensional peace operations) and "resident coordinators" (for non-conflict zone situations). Nevertheless, efforts through the *HDR* series and elsewhere to approach UN operations—and the promotion of "sustainable human development"—in a truly integrated manner continue to come up short.

For instance, while useful for coordinating the various components of UN programs in sometimes complex and sizeable field operations, the UN Development Assistance Framework (UNDAF) fails to show how the sector-specific components contribute to a coherent strategy for human development in a host country or region.[34] Part of the explanation for this is that the *HDR* has yet to extend beyond UNDP and, as envisioned by its chief architect, the late Mahbub ul Haq, be embraced as the flagship annual report by the UN family as a whole. Until this happens, and regular "core team" contributors are sourced from across the entire UN system, the continuation of somewhat rigid silos can be expected among the UN's various professional communities focused on peace and security, economic and social development, human rights and democratic governance, and environmental sustainability.[35]

Nevertheless, by instilling in UN civil servants the courage to think more independently, if not more rigorously and certainly more creatively, the *HDR*s contributed to a steady shift in the institutional culture of the UN system. Although few, if any, UN publications have yet to match the level of autonomy enjoyed by officers in UNDP's *HDR* Office, staff across the UN system are less restrained compared with previous decades in openly criticizing governments, multinational corporations, and other influential global governance actors. They are also more open to considering often far-reaching reform proposals rooted in sound research. For example, the *Human Development Report 1994* called for a new "United Nations Human Development Umbrella," which would allow for additional resources, responsibilities, and coordination among the UN programs, funds, and agencies.[36] Alongside the abovementioned Economic Security Council proposal, the idea to bring together the fragmented UN system around a common platform was viewed by many UN staff as useful to engender greater development resources, reduced duplication, and a more professional staff.[37] In short, over the past two and a half decades, the concepts, measurement tools, and even many of the reform proposals advanced through the *HDR*s, have begun to be internalized by officers within the UNDP and wider UN system. As noted above, especially in dismantling the barriers that preclude the adoption of truly integrated approaches to human development, further work remains if UNDP's "crown jewel" is to maximize its impact in fully transforming the UN's institutional culture.

Challenging the Washington Consensus: human development and international policy dialogues beyond the UN proper

During the 1990s, the influence of the human development approach was not limited to the UN proper, headquartered mainly in New York and Geneva, and the world body's series of global conferences. Rather, even in their early years, the *HDR*s would soon impact other major international policy dialogues, including in the international financial institutions, the WTO, the OECD, and the G7 (later G8). In particular, the *HDR*s made significant headway in articulating a coherent response and intellectual counterweight to the highly regarded "Washington Consensus" set of policy priorities advanced by key institutions of global governance, such as the World Bank and IMF.

Coined by the economist John Williamson in 1989, the term Washington Consensus was initially associated with a concrete set of ten economic policy reforms advocated by the World Bank and IMF for developing countries facing an economic or financial crisis. It encompassed policies ranging from trade and investment liberalization to macroeconomic stabilization. Soon after the Washington Consensus notion was introduced by Williamson, it took on a broader definition tied to those individuals and institutions vigorously espousing principles of "market fundamentalism" or "neoliberalism." The stage was set for a battle of ideas with the human development approach, which, from its very outset, expressed deep skepticism in unbridled capitalism that did not promote broad-based development—while quick to acknowledge the ample shortcomings of the state in the efficient allocation of economic resources. In contrast with a human development perspective that "emphasizes the *potential* benefits of the open global economy,"[38] Richard Jolly argues that "[the neoliberal approach] … underscores opening international markets, removing all barriers to trade and capital flows, and offering only the poorest countries some aid for a limited period."[39] In short, as depicted in Table 2.1, Jolly sums up the chief difference between human development and neoliberalism as their respective, predominant focus on poverty reduction and growth.

Sometimes described as the analytical underpinnings of a "New York Consensus" opposed to central tenets of an evolving "Washington Consensus," the *HDR*s sought to strike a more appropriate balance in society between the state and market forces. With their emphasis on public sector investments in education, health, and other

Table 2.1 Priorities compared: poverty reduction and growth

Human development	*Neoliberalism*
Key assumption: Growth must be consciously made pro-people and pro-poor	*Key assumption*: Trickle-down can be expected
Goal-oriented poverty strategy: • Empower the poor • Aim for gender equity • Ensure poor have access to assets • Accelerate pro-poor growth • International support for national action	*Growth-oriented poverty strategy*: • Ensure adequate economic growth • Expand social sectors • Build in safety nets as affordable • Open economy policies and international aid

Source: Richard Jolly, in *Readings in Human Development: Concepts, Measures and Policies for a Development Paradigm*, ed. Sakiko Fukuda-Parr and A.K. Shiva Kumar (New Delhi: Oxford University Press, 2003), 111.

essential areas for strengthening human capabilities, the earliest *HDR*s sought to reverse the wide intellectual pendulum swing toward markets in the 1980s and to restore faith in the fundamental governance functions of state institutions. One major example of the human development approach's growing resonance in the 1990s was manifested in the World Bank's *World Development Report 1997: The State in a Changing World*. Rather than concluding that the market role of states should be shrunk because of past failed interventions, *WDR 1997* argues that the state's role in the institutional environment underlying the economy (i.e., its ability to enforce the rule of law to underpin transactions) is vital to making government contribute more effectively to development.[40] However, the emphasis here is primarily economic governance functions, instead of privileging or at least affording equal weight to the—sometimes costly yet vital, as noted in the *HDR*s—social development and political freedom promoting functions of the state.

Another way human development challenged the Washington Consensus orthodoxy was by entering the lexicon of World Bank officials in the 1990s. For example, during the early 2000s, a new "network" for human development was established as the World Bank's home for policy, programs, and research in the fields of education, health, and social protection and labor: the World Bank Human Development Network (HDN).[41] Albeit adopting a significantly narrower definition of human development than UNDP and *HDR* authors, the creation of, for example, vice president, chief economist, and human development officer positions dedicated to human development has contributed to increasing concern for poverty reduction and strengthening human capabilities at the World Bank. And it is no coincidence that the World Bank shifted away from a core focus on top-down structural adjustment programs to more participatory *Poverty Reduction Strategy Papers* (*PRSP*s), between the 1980s and the early 2000s, as interest in the concept of human development moved to the fore internationally. Unlike structural adjustment, *PRSP*s place a premium on people's participation and the centrality of human agency for tackling the causes of human poverty, common themes in all *HDR*s, although perhaps amplified in the early years in the 1993 and 1997 annual reports.

Beginning with the *Human Development Report 1992*, the *HDR*s provided staunch advocacy for a more inclusive and fairer global trading system. As a forerunner to the negotiations leading to the World Trade Organization's establishment on 1 January 1995, the 1992 report called for the expansion of the General Agreement on Tariffs and Trade (GATT) to include all countries in the world, as well as the negotiation and enforcement of clear and fair rules within the GATT to encompass such issues as antidumping, subsidies, safeguards (emergency measures against imports), and restrictions on foreign investors.[42] In arguing that restricted global markets in key agriculture and textile sectors, coupled with barriers to immigration, costs developing countries $500 billion a year (10 times what they receive in foreign assistance and more than six times what they spend on human development priorities), *HDR 1992* contributed to the chief aims of the WTO's current "Doha Development Round" of trade talks that commenced in 2001.[43] These and related points, including the underlying causes for the stalled Doha Round talks, are further elaborated in Chapter 4 on energy, the environment, and trade.

In addition to the international financial institutions and World Trade Organization, the human development approach from its earliest days has shaped international policy deliberations in the Organisation for Economic Co-operation and Development and G7 (later G8) group of industrialized nations. For instance, the head of the United Nations Children's Fund (UNICEF), Jim Grant, employed the *HDR*s in 1991 to pressure major bilateral donors represented on the OECD's Development Assistance Committee to begin reporting on the percentage of their aid allocated to human development priority areas.[44] Many core human development principles for improved aid effectiveness—including on national ownership and donor alignment behind a partner country's national development strategies, institutions, and procedures—found their way into the 2005 Paris Declaration on Aid Effectiveness, spearheaded by the OECD-DAC.[45] And as noted earlier in this chapter, the *HDR*s provided the intellectual impetus for the Millennium Development Goals (2000–15) that would soon guide the donor policies and priorities of both G7/8 industrialized countries and the wider OECD-DAC donor community. Moreover, the Economic Security Council proposal, introduced in the 1992, 1993, and 1994 *HDR*s, would later inform the membership size and substantive focus of the G20.

A "New Framework for Development Cooperation"

In the early to mid-1990s the *HDR* series termed the phrase "New Framework for Development Cooperation" to frame a collection of reform innovations, including those outlined in this chapter (e.g., the 20:20 Compact, a Global Peace Dividend, an Economic Security Council, strengthening the UN's development agencies, and reforming the Bretton Woods institutions, the World Trade Organization, OECD-DAC, and G8/G20), plus various other ambitious measures. Key principles on which this new framework was based—and how it extends beyond conventional thinking in both the Global North and Global South—include:

- Mutual interests, not unilateral concessions (or charity).
- Two-sided responsibility, not one-sided accusations.
- More equitable access to global opportunities, not massive transfers of financial resources.
- More open markets, not more managed markets.[46]

In more policy-specific terms, the New Framework for Development Cooperation sought to broaden development cooperation beyond its predominant focus on foreign aid to include trade, private investment, technology, debt payments, and labor flows. Through the 20:20 proposal and other measures, Mahbub ul Haq and other *HDR* collaborators sought to demonstrate that core human development priorities could be financed by reallocating priorities in existing developing country and foreign assistance budgets. They also sought to redress the imbalance between short-term emergency assistance and long-term development support. They further aimed to establish new mechanisms: (i) to facilitate payments by one country to another for services rendered; (ii) to facilitate compensation for damages when one country inflicts an economic injury on another; and (iii) to mobilize resources for global objectives critical to human survival.[47] And as noted earlier, they sought to make military spending and security sector reform more generally legitimate development issues.

Integral to initiating and sustaining these policies was the call for new institutions of global governance, particularly in the economic realm. At their basic core, the *HDR*s advocated for a new global governance architecture or framework to set enforceable international rules, redress widening disparities within and among nations, and promote "global public goods" while mitigating the effects of "global public bads." And they sought to generate new sources of revenue for this global agenda through, for example, demilitarization funds, tradable pollution permits, and a small fee placed on global foreign exchange movements.[48] These and related progressive, internationalist policy and institutional reform initiatives left an indelible mark on international policy discussions in the 1990s, particularly during the series of UN world conferences between 1990 and 1996.

At the Rio Earth Summit of 1992, human development was present in efforts—anchored around the concept of sustainable development—to reconcile the world's finite natural resources with the vast material requirements of present and future generations if human development opportunities were to expand steadily.[49] Similarly, human development insights and policy proposals were featured in negotiations on the action programs at subsequent UN world conferences, perhaps most noticeably in the World Summit for Social Development Programme of Action's 10 Commitments, the Copenhagen Declaration, and an unofficial Copenhagen Social Charter. However, as elaborated in subsequent chapters and particularly Chapter 6, effective follow-through to major international policy discussions in the 1990s—in terms of implementing these agendas for action and enacting strong successor policies—was often lacking, in part because of the inherent weaknesses in global governance institutions to either facilitate progress toward commitments—through financing, technical capacity-building, and other inducements—or enforce them (sanctions). In short, despite significant normative advances advocated by the *HDR*s, the continued over-concentration of authority and resources at the level of the nation-state, at the expense of both global and sub-national levels of governance, precluded timely and credible responses to looming threats on the horizon to the security and well-being of large cross-sections of the world's population.

Conclusion

While most of the abovementioned policies and global governance reform ideas never came to pass in their pure, original form, this chapter has sought to demonstrate how the earliest *Human Development Reports* exerted profound influence on developing thinking and altered policy outcomes, during the past two and a half decades, in UN and other international policy forums. It also considered the extent to which human development discourse (including specific human development themes and issues advanced) has affected the field-based programming of international agencies from across the UN system, including the international financial institutions, as well as beyond it, such as at the World Trade Organization, OECD-DAC, and G8/G20.

Moreover, though the *HDR*s strive to have an immediate global impact and shape international policy discourses in the short-term, what is the "global shelf-life" of a good idea? Historically, powerful concepts and ideas—for example, capitalism, communism, or inclusive, democratic governance—can sometimes take decades to spread and take root, often evolving along the way. In this sense, are the *Human Development Reports* and many of their political and institutional reform innovations simply "Ahead of the Curve," as suggested in one book title on the power of new ideas conceived within the United Nations?[50] Time will tell, and as recommended in Chapter 6, many of the better reform proposals should be revisited, revised, and enriched in future *HDR*s at global, regional, national, and sub-national levels of governance.

At the same time, the first decade of *Human Development Reports* made significant strides in redirecting the attention of both states and markets towards people as the chief ends, as well as means, of development. Defined as "a process of enlarging people's choices" in order to improve the human condition—which, over time, came to also be accepted as an expansion of human capabilities, an enhancement of freedoms, and a fulfillment of human rights—human development has placed people at the center of analyses and policy recommendations in international policy debates. In doing so, it has transformed how we look at and understand development at the start of the twenty-first century. A key reason for the near-term success of the *HDR*s is the series' introduction of unique composite measurement tools and evaluative techniques, a subject to which we now turn in Chapter 3.

Notes

1 UNDP, *Human Development Report 1990* (New York: Oxford University Press, 1990), 1.
2 Khadija Haq and Richard Jolly, "Global Development, Poverty Alleviation, and North-South Relations," in *Pioneering the Human Development Revolution: An Intellectual Biography of Mahbub ul Haq*, ed. Khadija Haq and Richard Ponzio (New Delhi: Oxford University Press, 2008), 75.
3 The major themes of the earliest *HDR*s were concept and measurement (1990), financing human development (1991), international dimensions of human development (1992), people's participation (1993), human security (1994), and valuing women's work (1995).
4 Other conceptual, policy, and institutional reform innovations presented in the early years of the *HDR* included the promotion of national plans for human development (*HDR 1990*), a reformed system of official development assistance and a global debt bargain (*HDR 1992*), measures to decentralize governance and expand community organizations to increase local decision-making (*HDR 1993*), and new ways to account for women's contributions in the informal sector (*HDR 1995*).
5 Mahbub ul Haq, *Reflections on Human Development* (New York: Oxford University Press, 1995).
6 UNDP, *Human Development Report 1994* (New York: Oxford University Press, 1994), 77.
7 Haq, *Reflections on Human Development*, 184.
8 UN, "Programme of Action of the World Summit for Social Development," UN doc. A/CONF.166/9, 14 March 1995, www.un-documents.net/poa-wssd.htm.
9 Khadija Haq and Richard Jolly, "Global Development, Poverty Alleviation, and North-South Relations," in *Pioneering the Human Development Revolution*, ed. Haq and Ponzio, 76.

10 UNDP, *Human Development Report 1990*, 76.
11 UNDP, *Human Development Report 1992* (New York: Oxford University Press, 1992), 89
12 UNDP, *Human Development Report 1994*, 59.
13 Specifically, it is estimated that global military spending dropped from US$1 trillion in 1987 to US$767 billion in 1994 (in 1991 prices). Haq, *Reflections on Human Development*, 126. It is important to note that kindred spirits from the peace community, as well as policy networks that preceded the *HDR* (e.g., the previously mentioned North-South Roundtable), can lay claim to having influenced reductions in military claims following years of steady engagement.
14 As presented in the "New Deal for Engagement in Fragile States": "By September 2012, a set of indicators for each goal will have been developed by fragile states and international partners, which will allow us to track progress at the global and the country levels. These indicators will combine objective measures with measures to understand the views of people on results achieved." See www.g7plus.org/en/new-deal-document.
15 Similar to the Global Demilitarization Fund's areas of emphasis, the UN Peacebuilding Fund channels financial support (normally through a UN agency or peace operation) to disarmament, demobilization, and reintegration (DDR) programs and encourages inclusive governance and a strong civil society. See United Nations Peacebuilding Fund, www.unpbf.org.
16 For example, in 2005, the OECD's Development Assistance Committee adopted guidelines for security sector reform and governance.
17 See, for example, Andrew Mack, "Global Political Violence: Explaining the Post-Cold War Decline," Coping with Crisis Working Paper Series, International Peace Academy, March 2007.
18 UNDP, *Human Development Report 1993* (New York: Oxford University Press, 1993), 2. *HDR 1994* would elaborate that "Human security means that people can exercise these [human development] choices safely and freely—and that they can be relatively confident that the opportunities they have today are not totally lost tomorrow." UNDP, *Human Development Report 1994*, 23.
19 For further details on the UN Human Security Trust Fund, see www.un.org/humansecurity/, accessed on 30 October 2015.
20 According to the *Human Development Report 2010:* "This concept of human security has since become central to several global initiatives [e.g., International Commission on Intervention and State Sovereignty; the 2003 Commission on Human Security; and the 2004 High-Level Panel on Threats, Challenges and Change], has been picked up by national governments [e.g., Canada, Japan, Norway, and Switzerland], and is reflected in the agendas and policy debate of regional intergovernmental organizations [e.g., the African Union, the European Union, the Association of Southeast Asian Nations, the Organization of American States, and the League of Arab States]." UNDP, *Human Development Report 2010* (New York: UNDP, 2010), 17.
21 UNDP, *Human Development Report 1992*.
22 UNDP, *Human Development Report 1994*, 84.
23 Ibid.

24 In his book chapter outlining the case for a UN Economic Security Council, Mahbub ul Haq writes, "HIV/AIDS already has cost the international community more than $240 billion in research, curative measures and lost productivity in the past decade. Any viable plan to control the spread of this deadly disease will require upstream investment in preventive health care and downstream investment in containment, cure and research." Mahbub ul Haq, *Reflections on Human Development*, 193.

25 The outcome document from this session is UN Security Council Resolution 1325, 31 October 2000.

26 Quote by William Draper in Craig Murphy, *The United Nations Development Programme: A Better Way?* (Cambridge: Cambridge University Press, 2006), 241. Draper pursued and then recruited Haq as his senior advisor in 1989, having been deeply impressed by his intellect on meeting him in the 1980s during a visit to Pakistan.

27 Interview with Inge Kaul on 26 January 2007.

28 Desmond McNeill, "Human Development: The Power of the Idea," *Journal of Human Development* 8, no. 1 (2007).

29 Sakiko Fukuda-Parr and Selim Jahan, "Haq, Mahbub ul (1934–1998)" in *The Elgar Companion to Development Studies*, ed. D.A. Clark (Cheltenham: Edward Elgar, 2006), 215.

30 Asun Lera St. Clair, *Poverty Conceptions in the United Nations Development Programme and the World Bank: Knowledge, Politics and Ethics*, unpublished doctoral dissertation, University of Bergen, 2013, 216.

31 Craig Murphy, *The United Nations Development Programme*, 245–6. As Murphy further writes, the *Human Development Reports* "helped turn UNDP from being an organization that increased the dependence of people on their governments to one that also helped to keep governments accountable to their citizens." Ibid., 242.

32 Tatiana Carayannis and Thomas Weiss, "Ideas Matter: Voices from the United Nations," *Forum for Development Studies* 32, no. 1 (2005): 255.

33 Richard Jolly, Louis Emmerij, and Thomas Weiss, *UN Ideas that Changed the World* (Bloomington: Indiana University Press, 2009), 194.

34 Ibid., 195.

35 One possible exception is found in attempts to operationalize both the integrative concepts of human security (introduced in global *HDR 1994*) and peacebuilding. Whether through the UN Human Security Trust Fund launched in 1999 or the UN Peacebuilding Fund established alongside the Peacebuilding Commission and Peacebuilding Support Office in 2005, funding guidelines—as well as the actual practice of these multidonor trust funds—purposely ensured that programs, funds, agencies, and major departments (e.g., UNDP, UNICEF, UNHCR, and DPA) across the UN system contributed to conflict management in fragile and conflict-affected countries and regions. Moreover, in certain cases (e.g., the development and promotion of integrated peacebuilding strategies, as well as for security sector reform efforts that engage both development actors and traditional security agencies), the Peacebuilding and Human Security Trust Funds supported directly comprehensive approaches to achieving "positive peace," in line with emerging human development orthodoxy.

36 UNDP, *Human Development Report 1994*, 83–4.

37 Albeit narrowly circumscribed, several of the components associated with the UN Development Umbrella proposal inspired efforts several years later to forge "One UN" in field operations, as outlined in the previous paragraph.

38 Jolly further writes in comparing the human development and neoliberal economic approaches: "By contrast, the human development perspective has a more structuralist view … [it] recognizes the need for special action if the poorer and weaker countries are to find opportunities to participate fully and effectively … [T]his objective requires action to ensure a level playing field and support to strengthen the negotiating position of poorer and weaker countries. It also entails more aid and special assistance for the least developed countries." Richard Jolly, "Human Development and Neo-liberalism: Paradigms Compared," *Readings in Human Development: Concepts, Measures and Policies for a Development Paradigm*, ed. Sakiko Fukuda-Parr and A.K. Shiva Kumar (New Delhi: Oxford University Press, 2003), 113.

39 Ibid., 113. As the *Human Development Report 2010* further states: "By the early 1990s, the Washington Consensus had attained near hegemony, and mainstream development thinking held that the best payoff would come from hewing to its key tenets of economic liberalization and deregulation … From the outset the *HDR* explicitly challenged this orthodoxy and established a tradition that would be applied to a range of issues important to development policy." UNDP, *Human Development Report 2010*, 15.

40 World Bank, *World Development Report 1997: The State in a Changing World* (Washington, DC: World Bank, 1997). *WDR 1997* further argues against reducing government to a minimalist state, explaining that development requires an effective state that plays a facilitator role in encouraging and complementing the activities of private businesses and individuals. It presents a state reform framework strategy: first, focus the state's activities to match its capabilities; and second, look for ways to improve the state's capability by re-invigorating public institutions.

41 According to the World Bank, "HDN's mission is to invest in creating equal opportunities for people to live healthy, productive lives, secure meaningful jobs, and protect themselves from crises. HDN takes a lifecycle and systems approach to help developing countries deliver equitable and effective health, education, and social protection services from pregnancy to old age. HDN works across all development sectors and with Ministries of Finance to demonstrate how these investments in people promote inclusive development, economic growth, and country competitiveness." See the World Bank, Human Development Network, http://web.worldbank.org/WBSITE/EXTERNAL/EXTABOUTUS/ORGANIZATION/EXTHDNETWORK/0,,menuPK:514432~pagePK:64158571~piPK:64158630~theSitePK:514426,00.html, accessed on 30 October 2015.

42 UNDP, *Human Development Report 1992*, 82.

43 Ibid., 5–6.

44 Letter from Jim Grant to Alexander R. Love, chairman of the OECD Development Assistance Committee, 6 August 1991.

45 OECD, "The Paris Declaration on Aid Effectiveness (2005) and the Accra Agenda for Action (2008)," www.oecd.org/dac/effectiveness/34428351.pdf.

46 Mahbub ul Haq, "The Human Development Paradigm," in *Readings in Human Development*, ed. Fukuda-Parr and Shiva Kumar, 27.

47 Mahbub ul Haq, *Reflections on Human Development*, 138.

48 The latter proposal, known as the "Tobin Tax" (named after the Nobel prize winning economist James Tobin), was estimated in 1994 to raise around $150 billion a year through a tax as small as 0.05 percent on the value of each foreign currency exchange transaction. UNDP, *Human Development Report 1994*, 69.

49 The *HDR*s for 1992, 1994, and 1998 each considered environmental security from different angles and recognized the inherent unfairness associated with environmental degradation. In particular, they remind policy-makers that the debate about what should be sustained is as important as how to sustain it. Moreover, a human development approach stresses that intragenerational equity is as important as intergenerational equity.

50 Louis Emmerij, Richard Jolly, and Thomas G. Weiss, *Ahead of the Curve? UN Ideas and Global Challenges* (Bloomington: Indiana University Press, 2001).

3 Human development measurement tools

Advantages and shortcomings

- **Evolution of human development measurement tools since 1990**
- **Elevating the value of human development in global governance through measurement innovations**
- **The SDGs and the future of human development measurement tools in global governance**
- **Conclusion**

From the inaugural edition of the global *Human Development Report* in 1990, human development proponents designed evaluative mechanisms to quantify social and economic progress and to compare social groups within and between countries over time. Primarily in the forum of composite indicators, they sought to respond to the constraints of Gross National Product (GNP) as a measure of human progress, which was viewed as one-dimensional, failing to recognize non-monetized activities, and reflecting market prices in solely monetary terms.[1] Responding to various discontents about GNP as a sufficient measure for reflecting human well-being, the introduction of the Human Development Index (HDI) in 1990 had an immediate impact on international policy debates about national and human progress. It soon became the most recognizable and eagerly awaited feature of the annual *HDR*.

As an alternative to GNP and its associated per-capita income measure, the HDI began as a simple aggregate of indicators reflecting three major components of human development: longevity, knowledge, and command over resources needed for a decent living or income. The strengths and weaknesses of the HDI as a measure of human progress in global policy debates are examined, as well as the value of related composite indices such as the Gender-related Development Index (GDI), the Gender Empowerment Measure (GEM), the Human Poverty Index (HPI), and the Political Freedom Index (PFI). In particular,

HDI considers the extent to which human development measurement tools are translated, for example, into revised national budgetary allocations in greater favor of human development policy priorities. The chapter concludes with a discussion on the future of human development composite measures in global governance, including their relationship to the Post-2015 Development Agenda and "Sustainable Development Goals" (SDGs).

Evolution of human development measurement tools since 1990

Since the earliest *HDR*, the human development school or approach has sought to push the frontiers of measurement. Serving as the report series' flagship composite indicator, the HDI innovated thinking about human progress. For starters, it critiqued national economic aggregate measures, such as GNP and Gross Domestic Product (GDP), as well as per capita income, which together failed to adequately account for the totality of what constitutes development in the twilight years of the twentieth century. In measuring the average achievements in a country in three basic dimensions of human development—a long and healthy life, access to knowledge, and a decent standard of living (elaborated further below)—the HDI, like other measures, also does not take into account the full meaning of development. However, it does represent a radically significant improvement over earlier economic-centric measures.[2]

Employing calculations based on data from leading international data agencies (including the International Monetary Fund and Organisation for Economic Co-operation and Development) and other credible data sources, the HDI is the most widely cited aspect of a global *HDR*, whether by international policy-makers, scholars, or journalists. Cross-country comparisons fuel vigorous debates within and between each of these groups on whose country is ahead, whose country is behind, and what the primary factors are that contribute to a country's overall ranking, particularly in relation to the performance of neighboring countries.[3] Besides the chief architect of the *HDR*s—Mahbub ul Haq—Amartya Sen, Sudhir Anand, and Meghnad Desai[4] were instrumental in the conceptualization of the HDI as a composite statistic of human well-being. Worried about the challenges of reflecting the full extent and complexity of human capabilities in a single index, Sen initially opposed the HDI, writing in later years:

> I did not, I must admit, initially see much merit in the HDI itself, which as it happens, I was privileged to help him devise. I had

> expressed to Mahbub considerable skepticism about trying to focus on a crude index of this kind, attempting to catch in one simple number the complex reality of human development and human deprivation. Why give prominence, it was natural to ask, to a crude summary index that could not begin to capture much of the rich information that makes the *Human Development Reports* so engaging and important?[5]

Sen eventually came around when Haq convinced him of the value of a single number for focusing the attention of international and national policy-makers, as well as shifting their attention away from narrow economic concerns to a broad focus on major factors contributing to and reflecting advances in human progress. Sen also appreciated that "you could not pick-and-choose" variables arbitrarily (e.g., choose health at the expense of education or economic factors) and that, while still crude, the HDI was a more comprehensive measure than GNP. For Haq, the one-dimensional GNP was severely constrained because it remained "silent about the distribution, character, or quality of economic growth."[6]

Over time, the HDI, as a composite statistic comprising life expectancy, education, and income indices, has been employed to rank countries into four tiers of human development (see Figure 3.1). Here, for the year 2013, one observes the very highest levels of human

Figure 3.1 Human development levels worldwide, 2013
Source: World map indicating the category of Human Development Index by country, based on 2013 data, published in UNDP, *Human Development Report 2014* (New York: UNDP, 2014).

development—as represented through the HDI—in the Western countries of North America, Western Europe, Japan, Australia, New Zealand, Chile, and Argentina. This is followed by high levels of human development in the countries of Eastern Europe and Former Soviet Union, as well as large parts of Latin America and the Middle East. China, India, and other parts of Asia and the Middle East are then categorized in the medium quartile of the HDI. And finally, many countries within Sub-Saharan Africa and Central America, but also several small island developing states, fall within the fourth and lowest category of the HDI. By 2009, the index ranked 182 countries, but with changes undertaken to improve it in 2010 (see below), the number of countries dropped to 169.[7]

From 1990 until 2010, the three pillars of the index were calculated as follows:

1 Life expectancy at birth (representing population health and longevity).
2 Knowledge and education, with two-thirds weighting given to the measurement of adult literacy rate and one-third weighting to the combined primary, secondary, and tertiary gross enrollment ratio.
3 Standard of living, calculated by the natural logarithm of gross domestic product per capita at purchasing power parity (PPP).

In analyzing trends between 1975 and 2004 (Figure 3.2), one observes a steady improvement in the HDI across most regions of the world, especially in East and South Asia but also the OECD countries, Latin America and the Caribbean, and the Arab States. Following initial setbacks to human development in Eastern Europe and the Former Soviet Union (Commonwealth of Independent States) in the early 1990s, these countries had begun to bounce back by the mid- to late-1990s. Rather disturbingly, starting at a low HDI level just north of 0.4 in 1975, Sub-Saharan Africa managed to climb, only incrementally, over the following decade, and then basically flat-line well short of an HDI score of 0.5 in the subsequent two decades. This is all the more startling given that South Asia, known for equally acute levels of poverty and deprivation in the mid-1970s (and a similar HDI starting point as Sub-Saharan Africa) managed to cross the 0.6 threshold within three decades.

Particularly given the high-profile nature of the HDI since the inception of the first *HDR*, many criticisms—some constructive and well-substantiated, others not—have been lodged against the index. Concerns have ranged, for example, from the need to eliminate the

Figure 3.2 HDI trends, 1975–2004
Source: UNDP, *Human Development Reports*. See http://hdr.undp.org/en/faq-page/human-development-index-hdi.

composite measure's relative maximum and minimum values, and improving the method employed for income adjustment, to the absence of basic information for each of the statistics cited in the composite measure.[8] In the index's early years, constructive critiques came, for example, from scholars such as Leen Boer, Ad Koekkoek, Allen Kelly, and V.V. Bhanoji Rao.[9]

Other scholars have claimed that the HDI assumes an ideological bias toward Western models of development, fails to adequately account for ecological considerations, and focuses exclusively on national performance and ranking. But the index's three pillars—life expectancy, knowledge and education, and standard of living—relate to development goals around which countries in the Global North and South have forged a global consensus, as manifested in the Millennium Development Goals (MDGs, 2000–15). Underscoring the importance of sustaining human development advances as measured, in particular, by the HDI, global *HDR*s on, for example, consumption (1999), water (2006), and climate change (2007/8) have accounted for environmental impact and its connection to human development. And while the HDI employed in the global *HDR* compares performance between countries, this has not precluded the introduction of nationally disaggregated HDIs within many of the 130-plus countries that have produced National *HDR*s (including northern industrialized countries, such as the United Kingdom and United States), as well as municipalities such as Rio de Janeiro.

Emphasizing legitimate worries about erroneous data in the health, education, and income statistics employed to construct the HDI, the economists Hendrik Wolff, Howard Chong, and Maximilian Auffhammer recommended in an influential 2010 paper that UNDP discontinue what it viewed as an arbitrary practice of placing countries—as presented in Figure 3.1—in the categories of "Very High," "High," "Medium," and "Low" human development.[10] Wolff, Chong, and Auffhammer were rightly concerned that erroneous data underlying the HDI could lead to misinformed decision-making by policy-makers, investors, and development practitioners. They attributed data error to key factors such as (1) the need for better and updated data, (2) faulty formula revisions, and (3) and the need to better classify a country's level of development through more precise thresholds.[11]

To address several valid criticisms, the *HDR* Office of UNDP, starting in the 1990s, sought to refine each of the main variables and to disaggregate the HDI and related composite indices to provide the "clearest reflection to societies of prevailing realities, highlighting the disparities in human achievement or deprivations in terms of regions, states, provinces, gender, races, ethnic groups, and the rural-urban divide."[12] And beginning with *Human Development Report 2010*, a new method for calculating the HDI was introduced that sought, in particular, to respond to the critique presented above by Wolff, Chong, and Auffhammer.[13] For this updated version, UNDP developed three new indices: a Life Expectancy Index (LEI), an Education Index (EI), and an Income Index. The EI combines a Mean Years of Schooling

Index (MYSI)—the years that a 25-year-old person has spent in school—with an Expected Years of Schooling Index (EYSI)—the years that a 5-year-old child will spend in being educated during his or her entire life. In short, the updated HDI represents the geometric mean of these three indices.

The year 2010 also saw the global *HDR* series introduce the Inequality-adjusted Human Development Index (IHDI).[14] By accounting for inequality too, the IHDI is viewed by UNDP as the actual level of human development, whereas the HDI represents the "potential" level of human development to be achieved—or, if inequality was completely absent, the maximum IHDI that could be achieved. After more than two decades of *HDR*s, this important innovation has helped UNDP's flagship report return to a fundamental concern for the principle of equity and the quality, rather than simply quantum, of economic growth.

Composite indices beyond the HDI

Beyond the HDI, the *HDR*s have introduced over the years other new measures to evaluate progress in empowering women, reducing poverty, and promoting political freedom. The five most significant are discussed in order below: the Gender-related Development Index, the Gender Empowerment Measure, the Gender Inequality Index, the Multidimensional Poverty Index, and the Political Freedom Index.

The Gender-related Development Index (GDI) was introduced in *Human Development Report 1995* and sought to provide a gender-sensitive dimension to the HDI. It measures the gender gap in three basic areas of human development. The first is health, measured by female and male life expectancy at birth. The second, education, consists of the female and male expected years of schooling for children, and female and male mean years of schooling for adults aged 25 years and older. The third, command over economic resources, is measured by female and male estimated earned income. The GDI employs the same methodology as the HDI. Revealing pervasive gender gaps in human development up to the present day, the GDI, at the global level, indicates on average a female HDI value that is about 8 percent lower than the male value, and significant disparities also exist across countries, regions, and groups.[15]

The Gender Empowerment Measure (GEM) was also an innovation of the *HDR 1995*. It was conceived of as an instrument for capturing inequality in women's opportunities—rather than their capabilities—in three key areas. The first, political participation and decision-making power, is determined by the proportion of seats held by women in national parliaments. The second, economic participation and decision-making power, is based on the percentage of women in economic decision-making positions, including administrative, managerial, professional, and technical occupations. The third, power over economic resources, is determined by the earned incomes of males versus females.

The Gender Inequality Index (GII) was introduced in *Human Development Report 2010* as a variation of the GEM. The GII focuses attention on women's disadvantages in society in connection with three distinct capabilities: reproductive health, political empowerment, and participation in the labor market. Ranging from 0 (indicating that women and men fare equally) to 1 (indicating that women fare as poorly as possible to men in all measured dimensions), the GII represents the loss in human development because of inequality between female and male achievements in the above three dimensions.

Alongside the GDI, the GEM was widely discussed in terms of composite indices at the Fourth World Conference on Women, held in September 1995 in Beijing. Policy-makers and civil society activists have subsequently employed the GDI, GEM, and then (beginning in 2010) the GII, drawing on analytical trends presented in global, regional, national and sub-national *HDR*s, in advocating for more gender equality and stronger political, social, and economic participation by women in society. Besides considering the relationship between seemingly disparate issues, such as reproductive health, labor rights, and women's political participation, they have highlighted the links between women's advancement and a nation's overall human development progress. One review of human development trends since 1970, for example, found that the most robust predictors of HDI growth were, in fact, the level of fertility and female schooling rates.[16]

The Human Poverty Index (HPI), later supplanted by the Multidimensional Poverty Index (MPI), was first presented in *Human Development Report 1997*. It was introduced to better ascertain the extent of deprivation in developed countries compared with the HDI. At the same time, it focused on deprivation in the three areas highlighted by the HDI: longevity, knowledge and education, and a decent standard of living. One key feature of the Human Poverty Index was to derive separate deprivation measures for developing countries (known as "HPI-1") and high-income OECD countries (known as "HPI-2") to better reflect differences both in socio-economic levels and the tools for assessing deprivation.

In 2010, the MPI would improve on and replace the HPI by employing micro-data from household surveys to pinpoint deprivations—at the level of the individual in the areas of health, education, and standard of living—and how they overlap. Although it covers the same three pillars found in the HDI and IHDI and is, hence, multidimensional in nature too, the MPI, unlike the HDI and IHDI, must employ indicators that all emanate from the same household survey. For the survey, each person in a given household is classified as poor or non-poor depending on the number of deprivations experienced within a particular household. It is a useful tool for policy-makers and practitioners, especially as it can be disaggregated by region, ethnicity, and other groupings, as well as by pillar/dimension.[17]

Shortly after the initial construction of the HDI, critics raised the missing dimension of political freedom, admittedly a central element of the human development conceptual framework.[18] However, when *HDR 1992* put forth a Political Freedom Index (PFI, building on an attempt to devise a Human Freedom Index for *HDR 1991*), several UN member states objected officially, claiming that UNDP could not publish such an index without a mandate to work on human rights.[19] The then-prime minister of Malaysia, Mahatir Mohamed, condemned the PFI as an instrument of the West that interprets secular freedom as respecting the rights of homosexuals.[20] Curiously, however, it was the United States and other industrial countries that objected loudest to UNDP's two attempts to create a disinterested, neutral assessment of freedom in countries.[21] Believing that indicators of political freedom would someday be fused with the more established HDI, Mahbub ul Haq accepted that further work was required to refine the PFI and discontinued it after *Human Development Report 1992*.[22]

By complementing the HDI and IHDI, the five composite indices discussed above have helped to present a more complete measure of progress within a society. To employ these tools, only basic statistical data and mathematical knowledge are required, and they are widely understood by international policy-makers, practitioners, and scholars. Albeit with varying degrees of success, these composite measures have addressed concerns that the HDI reflects too limited a number of variables for representing the rich, complex, and dynamic concept of human development.

Elevating the value of human development in global governance through measurement innovations

Whereas the previous section provides a basic overview of how human development measurement tools evolved over the past two and a half decades, the following offers a brief introduction of the impact achieved by human development measurement innovations, particularly in affecting how international policy-making assesses and advances progress within and between nations. Specifically, the human development measurement tools outlined above have made their mark in the following five ways.

Gauging and communicating progress

First, they have provided policy-makers, development practitioners, and scholars with the tools to gauge progress and then communicate the progress achieved, as a result of various kinds of international development cooperation. One of the perennial challenges of the international development community has been to demonstrate to often skeptical legislative branches in donor countries and an even less informed general public, that their country's support for international development is achieving tangible purported outcomes. Ever since the 2008/9 global financial crisis, whose repercussions are still being felt in much of the industrialized world as of the time of writing in early 2015, it has become even more difficult in some industrialized countries to build and sustain the case for foreign aid, technology transfers, preferential trade deals, and other forms of international cooperation. The HDI and other tools are helping to counter the narrative against foreign assistance and broader forms of international cooperation.

By continuously reinforcing the message that "development is working" and in the interests of all countries in an increasingly interconnected world, the skillful use of human development composite measures since the first global *HDR* in 1990 has enabled diplomats and other advocates of international development to strengthen their case for sustained support from donor countries and philanthropic institutions. For instance, one influential background paper for *Human Development Report 2010* found progress in raising HDI levels to be fastest in low-HDI and middle-HDI countries between 1970 and 1990, and that the story of a slow "social convergence" between rich and poor countries continues until 2005, particularly with respect to the life expectancy and literacy dimensions of the index.[23] Another insight from HDI trends since 1970 is that income and non-income components of the HDI have a near-zero correlation, which suggests that proper public (supported by foreign assistance) investments in health and education can make a difference even in countries with relatively low levels of per capita income.[24] Sakiko Fukuda-Parr, Kate Raworth, and A.K. Shiva Kumar further elaborate on this critical point for policy analysis and policy-making when they contend:[25]

> The HDI's global ranking of countries provides an assessment of a nation's average achievement in human development, comparing it with the progress of other countries … First, it is possible to have similar levels of human development (as captured by the HDI) but very different levels of per capita income … Second, it is possible to have similar levels of income but very different levels of human development. Third … a higher income does not by itself imply a higher level of human development. Fourth, it is not necessary for a country to become rich first before it can assure people a decent level of human development. Fifth, misplaced priorities, and not necessarily a shortage of resources, often prevent countries from assuring people universal access to basic health and education.

Over the years, to reinforce these points, diverse countries such as Costa Rica, Botswana, Vietnam, and even Sri Lanka during decades of conflict showed that significant improvements in HDI levels could be achieved through good domestic policies combined with international assistance, even with relatively moderate levels of economic growth. Besides undertaking the heavy lift in terms of statistical number crunching and analyzing trends over time for each of its comprehensive indices, another contribution of the *HDR* series has been the creative presentation and broad promotion of its analytical findings through global, regional, national, and sub-national reports, as well as other mediums. Investing in carefully crafted communication strategies—which employ, for example, tables, charts, graphs, and colorful maps to depict change over time in human development indicators—has also been a hallmark of the *HDR*s. For hardened technocrats and even politicians that only view a policy decision as real (and worthy of public financing) if the associated policy issue or phenomenon can be measured, visually attractive presentations of the HDI and other composite measures have served as central elements of the communications toolbox employed by UNDP worldwide and partners in the wider international development community.

Informing international policy-makers

Second, human development measurement tools have informed international policy-makers about gaps in progress and where additional international assistance is required. Although overall trends in human development are positive globally,[26] many countries continue to fall short of these trends and suffer human development setbacks because of a variety of factors, such as poor governance, war, natural disasters, and continued "horizontal inequalities" among various types of groups (e.g., ethnic, religious, and social castes) within a society. The human development composite measures have helped both government and civil society groups analyze gaps in human progress to determine where the right kinds of public policy and foreign assistance can help to reduce these gaps.

According to Selim Jahan,[27] "[T]he HDI has served as a powerful instrument of public communication, playing a strong advocacy role and contributing significantly to policy debates and dialogues … The regional disparities revealed by the disaggregated HDI in Brazil's *National Human Development Reports* have prompted civil society institutions to highlight the issue of inequality and demand measures to reduce it." They have served, in the view of Thomas Weiss, as one of the most powerful tools of accountability available for intergovernmental organizations (IGOs) and international nongovernmental organizations (NGOs) to embarrass governments that are underperforming.[28] Viewing human development composite indicators as "a process that constitutes a major audit on implementation to help plug the compliance gaps," Weiss argues that "Defensive reactions suggest that embarrassment makes a difference; while it may not always improve compliance, sometimes it does."[29] Besides IGOs and international NGOs, former prime ministers also have noted the HDI's usefulness. For example, India's I.K. Gujral is on record as stressing the utility of referring to sliding human development indicators to ensure greater accountability and performance by individual Indian state government heads.[30]

The HDI and other human development indices also have been employed for analyzing development gaps in the nearly one-third of countries and territories classified by the World Bank and OECD as "fragile and conflict-affected,"[31] which overlap more or less with what the economist Paul Collier labels the "bottom billion" of the world's poor, who inhabit these polities.[32] The *Kosovo Human Development Report 2004*, for instance, informed policy-making by the then-Provisional Institutions of Self-Government and large UN Mission in Kosovo (UNMIK) by exhibiting the disparities in HDI levels between municipalities across the territory of Kosovo and recommending a variety of policy responses. Besides gaps between states, disaggregated human development data at different levels of governance within a state and between specific groups in a society have equipped decision-makers with the ability to identify and respond to development gaps before they become strikingly large and potentially destabilizing (see the next two points below).

Analysis and policy-making

Third, human development composite measures are employed for analysis and policy-making at various levels of governance. The HDI is famous for comparing across nations the aggregate measure of a country's progress within the three pillars of life expectancy, literacy, and school enrollment (the knowledge and education pillar), and income per capita (the standard of living pillar). Meanwhile the HDI and other human development composite measures also have aided international policy-makers, development practitioners, and scholars in assessing the estimated aggregate progress of the world, specific regions, and within sub-national units. This includes the application of innovative measurement methodologies to the comparison of neighborhoods within large municipalities in the Global North and South, including New York City and Rio de Janeiro. Urban-rural differentiation within countries has also aided decision-makers in understanding better the extent to which and why certain parts of a country or subregion are being left behind despite human development progress within a national or sub-national political unit.

Shedding light on domestic group distinctions

Fourth, indices are disaggregated to shed light on differences between specific kinds of groups in a particular polity. In addition to the geography, each human development composite indicator can also be disaggregated along other lines, including male/female, age group, income level, and ethnic group. In doing so, they further advance thinking and assessments of the rich and multifaceted concept of human development among government and donor policy-makers. Moreover, these types of disaggregation have proven instrumental in helping decision-makers determine priorities and formulate human-development-related policies and reforms that aim to bridge the kinds of inequality gaps that if left unchecked may continue to fester and grow, and can prove politically de-stabilizing and engender deadly violence and material destruction.[33]

Informing development policy debates

Fifth and finally, these indices have informed debates on future global development priorities. Beginning in the early to mid-1990s, the data collected, assembled in composite indicators, and analyzed had an immediate impact in informing and shaping global development priorities agreed at a string of UN world conferences. Beginning with the UN Conference on Environment and Development (Earth Summit) in 1992 in Rio de Janeiro, these meetings continued with the UN Conference on Human Rights in 1993 in Vienna, the UN Conference on Population and Development in 1994 in Cairo, the World Summit on Social Development in 1995 in Copenhagen, the Fourth World Conference on Women in 1995 in Beijing, and the UN Conference on Human Settlements in 1996 in Istanbul. Recognizing that growing GNP and per capita income in a society were necessary yet insufficient indicators of broader human progress, participants in these conferences unpacked the pillars of the HDI and GII composite indices to arrive at a more complete set of policy measures that were later enshrined in a conference's concluding program of action. As noted above, policy-makers and civil society activists attending the Fourth World Conference on Women gave attention to the analysis and policy recommendations flowing from the innovative GDI and GEM. In particular, the GDI and GEM generated a high-profile policy discussion in Beijing on the under-valued and often misunderstood contribution of women in their families, communities, and societies, measured in more holistic, human development terms.

In restoring a balance between human development priorities, such as education, health care, gender empowerment, potable water, and sanitation, and the earlier privileged status of economic growth strategies, the *HDR*s and their associated composite measures fed into the policy dialogue culminating, in September 2000, in the leaders of 189 countries pledging to make progress on eight MDGs over 15 years.[34] Alongside sector-specific statistics associated with targets for each of the eight MDGs, the comprehensive indices of the *HDR*s have contributed to periodic progress assessments; the entire *HDR 2004*, in fact, was dedicated to an update on the MDGs and how progress toward their achievement contributes to gains within the broader framework of human development. Although progress has been achieved in most of the MDGs, shortfalls are identified in critical areas such as education worldwide and environmental sustainability.

Given these gaps and fears of growing inequalities following the 2008/9 global financial crisis, plus the recognition of the value of the MDGs for focusing government leaders in developing countries and their donor partners in developed countries, world leaders are now preparing to commit themselves, in September 2015 in New York, to a new set of Sustainable Development Goals (SDGs). Here again, recent innovations in human development composite indices, including the

IHDI and GII, are once again expected to prove valuable in shaping priorities associated with the Post-2015 Development Agenda.

In the five areas presented above,[35] the human development measurement tools introduced in recent decades have made a profound impact in shaping how international policy-makers, development practitioners, and scholars view and attempt to promote international development. In short, they have provided new analytical tools to reinforce human development policy priorities and the broader paradigm shift away from the focus of many in the development community on national-income accounting and economic growth strategies. Efforts to strengthen the methodology of human development composite measures, as highlighted in the previous section, allowed the *HDR*s to offer a more objective and candid analysis of a country's development successes and failures. This, in turn, enhanced the credibility of their findings, elevating the *HDR*'s status as an international and national advocacy and policy-making tool.

Even with the methodological refinements made to several of the composite indices in recent years, Mahbub ul Haq, Amartya Sen, and other architects of the original instruments for gauging human progress would probably conclude today that neither the HDI, nor other composite measures, represent the totality of what constitutes human development. Nevertheless, they would continue to view these tools as vast improvements over a more narrowly focused, and even more crude, preoccupation with GNP and per capita income as the chief measures of human progress.[36] The HDI, in particular, remains the most widely cited aspect by far of the United Nations system's most popular annual report, and it is the only composite measure to be published by an intergovernmental organization. Provided that interesting and provocative thematic debates will also continue to be introduced in the annual global *HDR*, one can presume that the popularity of the HDI and associated composite measures will continue.

The SDGs and the future of human development measurement tools in global governance

Since the introduction of the original HDI in the *Human Development Report 1990*, human development composite measures have come a long way in influencing how people think about and assess human progress over time. To continue to impact international development debates, they will need to remain responsive to supporting current global development priorities, as well as being forward-leaning tools that continue to push the boundaries of how development is conceived and tracked in the twenty-first century.

In the former category, considerable political and analytical energy will be expended, for the period 2015 to 2030, on the Post-2015 Development Agenda—the successor to the MDGs introduced in 2000. As the High-Level Panel of Eminent Persons on the Post-2015 Development Agenda stated in its seminal 2013 report, "A New Global Partnership: Eradicate Poverty and Transform Economies through Sustainable Development":

> So a new development agenda should carry forward the spirit of the Millennium Declaration and the best of the MDGs, with a practical focus on things like poverty, hunger, water, sanitation, education and healthcare. But to fulfil our vision of promoting sustainable development, we must go beyond the MDGs. They did not focus enough on reaching the very poorest and most excluded people. They were silent on the devastating effects of conflict and violence on development. The importance to development of good governance and institutions that guarantee the rule of law, free speech and open and accountable government was not included, nor the need for inclusive growth to provide jobs. Most seriously, the MDGs fell short by not integrating the economic, social, and environmental aspects of sustainable development as envisaged in the Millennium Declaration, and by not addressing the need to promote sustainable patterns of consumption and production. The result was that environment and development were never properly brought together.[37]

At the time of writing, UN member states had coalesced around 17 SDGs as successors to the current eight MDGs. Although NGOs and several, especially Northern industrialized, countries have lobbied hard for the inclusion of at least one goal related to fundamental governance, justice, and security themes, many detractors are actively voicing opposition to this proposal. Particularly in the Global South, countries are concerned that an SDG on sensitive issues of national policy-making and internal security could open the door for greater outside (i.e., Western) interference in what they view as essentially the sovereign, internal affairs of a state. Major emerging economies, such as Brazil and South Africa, have also argued that if certain countries wish to emphasize governance reforms, they would be wise also to give attention to inequities and inefficiencies in the global system of governance.

However the Post-2015 Development Agenda debate plays out, human development measurement tools will have a valuable role to assume in tracking progress and identifying compliance gaps towards reaching the specific targets associated with the SDGs agreed in September 2015. In this regard, and recognizing that environmental issues are expected to assume a far more central place in the SDGs for 2015–30 compared with the MDGs for 2000–15, proposed new and/or additional modifications to human development composite indices, ranging from the HDI and MPI to the various gender-sensitive human development measure tools, include "Resource foot-printing analysis," "green accounting," and supporting the focus of SDGs on delivering access to basic services to all. Each is discussed in turn below.

First, resource foot-printing analysis could combine several dimensions (e.g., water, energy, and carbon footprints) but distinguish the impacts through alternative measurements. For instance, a production-focused approach to economic development might find large resource footprints in emerging economies, particularly China. But a consumption-focused approach would draw attention to the highly resource-intensive consumption patterns and lifestyles in developed economies. The purpose of such analysis would be to draw attention to multiple SDG goals (economic development, employment, environmental sustainability, etc.) rather than celebrate successes or shame failures on individual indicators.

Second, green accounting and accounting for the value of ecosystem services has been suggested for some time. But these are difficult to do unless certain baselines are established and new cost-benefit frameworks are developed. For better or for worse, the human development composite indicators helped to demonstrate correlations between core development priorities. Similarly, new accounting standards would have to be developed, which gives countries a framework to assess how their choices impact other development outcomes (say, investments in particular kinds of energy sources could have overall negative consequences for public health, natural capital depletion, etc.). The purpose of new accounting methods would not be to impose one solution on every country, but to offer a toolkit for more informed decision-making.

A third way in which new indicators would be helpful is in supporting the focus of SDGs on delivering access to basic services to all. This focus on reducing inequalities within countries (and not merely measuring aggregate achievements at a country level) would likely correct for a major flaw in how the MDGs were designed. It would also endorse the human development approach by focusing on the entitlements and capabilities of each individual.

Sustainable energy for all, for example, is one such priority. Once again, new indicators are needed to assess what energy poverty is, how the metrics of such deprivation change over time, the conditions under which sustainable energy sources are adopted by poor households, and the economic, social, and environmental impacts of such choices. Drawing on a large survey of 8,566 households in 714 villages across India, scholars from the Council on Energy, Environment and Water and Columbia University have undertaken one such attempt in the pursuit of developing new indices with relation to access to electricity and access to modern forms of cooking energy.[38] Similar access-related indices should be constructed for other services, such as clean water, modern sanitation, access to clean air and toxin-free environments, and so forth. Here, the indicators would consider increased morbidity and higher mortality rates, as well as economic losses from the absence of such basic services for all.

The above suggestions are merely indicative, and the list of new composite indicators might be longer and/or more refined. However, the development of composite indicators to align with the SDGs would build on the human development approach in three ways. Firstly, they would again focus on the individual or the household as the core unit of analysis. Secondly, they would shift attention away from individual indicators and, instead, establish the core relationships between various development goals (including tradeoffs between them). As a result, thirdly, new composite indicators could become the basis of bringing in a "data revolution"—as called for in the recent UN secretary-general's report on "A Life of Dignity for All"[39]—and serve as new frameworks for reigniting international development cooperation.

The missing dimensions of human development measurement tools

Particularly if at least one goal related to governance, justice, and security themes is included in the final set of SDGs, when world leaders gather in September 2015 in New York, a separate, forward-leading modification would merit consideration. This is the, in many ways, overdue integration of political, human rights, and cultural liberty dimensions of human development within the HDI and other human development composite measures. Learning from the political and methodological failures associated with introducing a Human Freedom Index in *HDR 1991* and Political Freedom Index in *HDR 1992*, the time has come to revisit Mahbub ul Haq's idea from the early 1990s that the political dimensions of human development could someday be fused with and accounted for within the HDI.

As noted in the first section of this chapter, both the HFI and PFI failed to establish objective and widely acceptable political freedom indicators for which data were widely available and for which change over time could be measured. In short, Haq opted to discontinue them because of an acknowledged lack of methodological rigor, especially compared with what had already been achieved with the HDI. But as the solid and credible data and new indicators—albeit qualitative and more subjective in nature compared with the quantitative data gathered for the current three socio-economic pillars of the HDI—demonstrate, the conditions have significantly improved for overcoming past methodological hurdles. These were assembled for the *HDR*s on human rights (2000), democracy (2002), and cultural liberties (2004).[40] Especially as attention will turn back to key national and global governance questions in 2016 on how to best deliver the new Post-2015 Development Agenda, the time is ripe to entertain Mahbub ul Haq's wish shortly following the inception of the Post-2015 Development Agenda.

Conclusion

From the outset of the global *HDR* series in 1990, the HDI (which compares progress between countries but also sub-national, regional, and global units of governance) played an influential role in shaping how international policy-makers, practitioners, and scholars understood and approached international development. Given their expansive scope—far beyond the initial three-pillar focus of the original HDI—other innovative human development measurement tools, introduced over the next two decades, would further enrich the influence of the integrative concept of human development. A concept which is, in essence, about the process of enlarging people's choices and building human capabilities or agency to realize those choices. In this sense, the human development composite measures made marked contributions to the first two of this project's four primary research questions, namely on the promotion of a paradigm shift and agenda-setting in key global institutions.

On the shaping of "new horizon issues" and overcoming "obstacles to progress" (the study's third and fourth primary research questions), refinements in the HDI and MPI, as well as creation of a new GII, have addressed key technical challenges and kept the comprehensive human development measurement tools relevant in gauging important international development trends in the early twenty-first century. In particular, they have contributed to both defining and assessing progress on the MDGs, and with suggested further modifications and improvements, as outlined above, the human development measurement toolbox is poised to assist the tracking of the post-2015 SDGs too. Similar to focusing on a defined set of goals associated with the MDGs and SDGs, Chapters 4 and 5 delve deeply into a select number of thematic issues for which the human development approach has made a significant mark in global governance.

Notes

1 Richard Ponzio, "The Advent of the *Human Development Report*," in *Pioneering the Human Development Revolution: An Intellectual Biography of Mahbub ul Haq*, ed. Khadija Haq and Richard Ponzio (New Delhi: Oxford University Press, 2008), 100.

2 As I argued earlier, "The HDI has helped policymakers focus on key objectives of development (longevity and knowledge), rather than the means (income); further it has proved to be a more meaningful national average than GNP as income distribution varies far more greatly within a country than the distribution of life expectancy and literacy." Ponzio, "The Advent of the *Human Development Report*," 102.

3 According to Selim Jahan, the current director of UNDP's Human Development Report Office, "the HDI has motivated desirable and healthy competition among countries to surpass neighbors or competitors in rankings." Selim Jahan, "Evolution of the Human Development Index," in *Readings in Human Development: Concepts, Measures and Policies for a Development Paradigm*, ed. Sakiko Fukuda-Parr and A.K. Shiva Kumar (New Delhi: Oxford University Press, 2003), 133.

4 The HDI's earliest conceptual work drew on a study by Meghnad Desai, in 1989, for UNDP's Regional Bureau for Latin America and the Caribbean on alternative measures of well-being to GNP.

5 Amartya Sen, "Mahbub ul Haq: The Courage and Creativity of His Ideas," speech at the Memorial Meeting for Mahbub ul Haq at the United Nations. New York, 15 October 1998.

6 Mahbub ul Haq, *Reflections on Human Development* (New York: Oxford University Press, 1995), 46.

7 David A. Hastings, "A 'Classic' Human Development Index with 232 Countries," Human Security Index, 2011, www.humansecurityindex.org/?page_id=204.

8 Ranging in opinion from broad agreement to pointed criticism, by 2003 over 40 academic articles had been written on the Human Development Index. See Kate Raworth and David Stewart, "Critiques of the Human Development Index: A Review," in *Readings in Human Development*, ed. Fukuda-Parr and Shiva Kumar, 164–7.

9 Leen Boer and Ad Koekkoek, "Human Development Report: Fad or Fixture?" *Development Policy Review* 11, no. 4 (1993): 427–38; Allen Kelley, "The Human Development Index: Handle with Care," *Population and Development Review* 17, no. 2 (2004): 315–24; and V.V. Bhanoji Rao, "Human Development Report 1990: Review and Assessment," *World Development* 19, no. 10 (1991): 1451–60.

10 Hendrik Wolff, Howard Chong, and Maximilian Auffhammer, "Classification, Detection, and Consequences of Data Error: Evidence from the Human Development Index," National Bureau of Economic Research Working Paper No. 16572, December 2010, www.nber.org/papers/w16572.
11 Ibid., 1.
12 Selim Jahan, "Evolution of the Human Development Index," in *Readings in Human Development*, ed. Fukuda-Parr and Shiva Kumar, 160.
13 UNDP, *Human Development Report 2010* (New York: UNDP, 2010). In a comment to *The Economist* in early January 2011, UNDP responded further to the paper by Wolff, Chong, and Auffhammer by arguing that, beginning with *HDR 2010*, it undertook a systematic updating of methods which ensures a continuous revision of human development categories whenever formula or data (that underlie the HDI's three pillars) are revised.
14 UNDP, *Human Development Report 2010*, 87. Note that the indicators needed to construct the IHDI need not come from the same survey.
15 See United Nations Development Programme *HDR*s, Gender-related Development Index, hdr.undp.org/en/content/gender-development-index-gdi. Note that across countries, gender gaps in HDI values range between 0 percent and 40 percent.
16 George Gray Molina and Mark Purser, "Human Development Trends since 1970: A Social Convergence Story," Human Development Research Paper 2010/02, UN Development Program, New York, June 2010, 27–8.
17 Building on advances in theory and available data, the MPI offers a valuable complement to other income-based poverty measures. *HDR 2013*, for example, presents estimates for 104 countries with a combined population of 5.4 billion (or 76 percent of the world total). The MPI for these countries indicates that a total of 1.6 billion people—or 30 percent of their entire population—could be classified as living in multidimensional poverty during the period 2002 and 2011. UNDP, *Human Development Report 2013* (New York: UNDP, 2013).
18 According to Meghnad Desai, efforts to devise a Political Freedom Index came about because the United States was unhappy with its HDI ranking and complained that many socialist countries ranked high on the HDI because the index ignored freedom. Interview with Meghnad Desai, 4 May 2006.
19 Meghnad Desai, "Measuring Political Freedom," *Readings in Human Development*, ed. Fukuda-Parr and Shiva Kumar, 196.
20 "So he denounced our index which did not include the article [about sexual freedom in the UN Declaration of Human Rights] as it was built on 15 articles only and not the entire UNDHR, saying he would not expose young boys to pedophiles just to satisfy UNDP's human rights measure," recounts Meghnad Desai. Interview, 4 May 2007.
21 Interview with Inge Kaul, 26 January 2007.
22 Haq, *Reflections on Human Development*, 72.
23 Gray Molina and Purser, "Human Development Trends since 1970," 26.
24 Ibid., 26–8.
25 Sakiko Fukuda-Parr, Kate Raworth, and A.K. Shiva Kumar, "Using the HDI for Policy Analysis," in *Readings in Human Development*, ed. Fukuda-Parr and Shiva Kumar, 153–4.

26 For example, the world reached the target of halving extreme poverty by 50 percent by the year 2010, five years prior to the ambitious 2015 date set by world leaders for achieving the 15-year Millennium Development Goals (this was primarily attributed to the tremendous socio-economic advances achieved over the previous decade by the world's two largest countries, China and India).

27 Selim Jahan, "Evolution of the Human Development Index," in *Readings in Human Development*, ed. Fukuda-Parr and Shiva Kumar, 134.

28 Thomas G. Weiss, *Global Governance: Why? What? Whither?* (Cambridge, Mass.: Polity Press, 2013).

29 Ibid., 176.

30 Craig N. Murphy, *The United Nations Development Programme: A Better Way?* (Cambridge: Cambridge University Press, 2006), 256.

31 Scholars at The Brookings Institution predicted in 2011 that more than 50 percent of the world's countries would be classified as "fragile" by 2015, up from only 20 percent in 2005. See Laurence Chandy and Geoffrey Gertz, "Poverty in Numbers: The Changing State of Global Poverty from 2005 to 2015," Policy Brief 2011–01, The Brookings Institution, Washington, DC, 2011, 10.

32 Paul Collier, *The Bottom Billion: Why the Poorest Countries are Failing and What Can be Done About It* (Oxford: Oxford University Press, 2007). Today, in both fragile and non-fragile states combined, 1.2 billion people continue to live under the extreme-poverty rate, and one out of every eight persons continued to suffer from hunger. Both targets are integral elements of the United Nations' first Millennium Development Goal.

33 For a detailed introduction to how growing "horizontal inequalities" between groups of various kinds can precipitate armed conflict, see Frances Stewart, "Horizontal Inequalities: A Neglected Dimension of Development," Queen Elisabeth House (University of Oxford) Working Paper Series No. 81, 2002. A celebrated professor of development economics at the University of Oxford, Frances Stewart worked with Mahbub ul Haq in the 1970s at the World Bank and, along with Amartya Sen, Paul Streeten, Gustav Ranis, and Meghnad Desai, was part of the original team that conceived of and developed the earliest *HDR*s.

34 As the old proverb goes, "success has many fathers, failure is an orphan." The *HDR* team claimed credit for influencing policy debates among OECD donor countries in the mid- to late-1990s on development targets that would lay the foundations for the Millennium Development Goals. In addition, the secretariats of the OECD and several specific initiatives (e.g., through UNESCO and the World Conference on Education for All, held in 1990 in Jomtien, Thailand) would also share in the credit for shaping what has, arguably, been the most significant set of development priorities since the turn of the century.

35 Beyond a sound methodology, Sakiko Fukuda-Parr, Kate Raworth, and A.K. Shiva Kumar further base the HDI's success on two other factors—policy relevance and acceptability—when they write, "First, policy-makers have found it useful and wanted to see it continue. Second, they have accepted it even if they have not always liked the results." They attribute the HDI's policy relevance and acceptability to four key features: conceptual clarity that facilitates the HDI's power as a tool of

communication; a reasonable level of aggregation; use of universal criteria amenable to intercountry comparisons; and use of standardized international data that have been legitimized through official processes. Fukuda-Parr, Raworth, and Shiva Kumar, "Using the HDI for Policy Analysis," in *Readings in Human Development*, ed. Fukuda-Parr and Shiva Kumar, 161.

36 Even scholars who have theoretical objections to composite indices often grant an exception to the HDI on the grounds that it has had a positive policy impact. For example, see Heba Handoussa, "Human Development Design of Indicators," discussion at the International Association for Official Statistics (IAOS) conference on "Statistics Development and Human Rights," session I-PL, 6/7, Montreux, 7 September 2000.

37 United Nations Secretary-General's High-Level Panel of Eminent Persons on the Post-2015 Development Agenda, "A New Global Partnership: Eradicate Poverty and Transform Economies through Sustainable Development," 2013, Executive Summary.

38 Abhishek Jain, Sudatta Ray, Karthik Ganesan, Michaël Aklin, Chao-yo Cheng, and Johannes Urpelainen, *Access to Clean Cooking and Electricity – Survey of States*, New Delhi: Council on Energy, Environment and Water (September, 2015)

39 UN Secretary-General, "A Life of Dignity for All: Accelerating Progress Towards the Millennium Development Goals and Advancing the United Nations Development Agenda beyond 2015," UN doc. A/68/202, 26 July 2013.

40 The highly respected work of UN human rights special rapporteurs and independent reporters could also be referenced and incorporated into these prospective new dimensions of human development composite measures. For an authoritative overview of the achievements of special rapporteurs and independent reporters in recent decades, see Ted Piccone, *Catalysts for Change: How the UN's Independent Experts Promote Human Rights* (Washington, DC: Brookings Institution Press, 2012).

4 Human development in international policy-making, Part I

Trade, water, energy, and environment

- **The relationship between sustainable development and human development**
- **Trade: with a human face**
- **Water: for life, for livelihoods, and for all**
- **Energy and environment: access not excess**
- **Conclusion**

In a typical election campaign in Hindi-speaking states of northern India, two slogans can be heard more than any other. Candidates promise, "*roti, kapda, makan*" (food, clothing, shelter). In places where some of these needs are already met, the campaigns emphasize more livelihood-related concerns, namely "*bijli, sadak, paani*" (electricity, roads, water). As the precursor to the human development approach, the emphasis on basic needs also highlights two sets of needs: for private consumption (food, shelter, and clothing), and for essential services (safe drinking water, sanitation, public transport, health, and education).[1] That the intellectual focus on basic needs is so similar to that during Indian election campaigning is no accident. They reflect a recognition of the core foundations for human development, in terms of building the capabilities that allow individuals to expand their life choices and lead long, healthy, and meaningful lives. They are also recognition of the political salience of these concerns in democratic societies.

Have these core foundations of human development—access to water, energy, a clean environment, and opportunities for trade and income generation—also penetrated international institutions, and if so how? Have they translated into policy approaches and innovations in global governance? Have they influenced the goals that international institutions set and the strategies they have adopted?

This chapter applies the lens of critical aspects of human development to explore its evolution in terms of the debates on trade, water, energy, and the environment. It first discusses the conceptual relationship between sustainable development and human development with particular reference to equity, responsibility, consumption, inter- and intragenerational sustainability, and the question of the ends and means of human development. This conceptual relationship has only grown in significance because of increasing concerns about the environment and introduction of the Post-2015 Development Agenda and Sustainable Development Goals.

The next section discusses the multilateral trade regime and how its legitimacy was questioned in the 1990s, even as it became more formalized under the WTO. It investigates how a human-centered approach shaped the discourse, particularly on the issue of access to medicines, and the rights of indigenous people and their knowledge. It also elaborates how the analysis in *Human Development Reports* (*HDR*s) and the strategic timing and targeting of such analysis helped to shape the agenda on aid directed to build capacity for trade in the least developed countries.

The following section discusses water, arguing that the human development approach reframed how deprivations in access to water and sanitation were understood. It then analyses two issues on which the human development approach had an influence, namely the International Decade for Action "Water for Life" and the tensions surrounding transboundary waters.

Finally, a section on energy and the environment brings the focus back to a wider understanding of sustainability. It describes the paradigm shifts that *HDR*s introduced on human security, impacts of environmental stresses on the poor, equity across geographies and between generations, and sustainable consumption. It then explores policy evolution in the UN system and international negotiations, with regards to energy access, technology and finance, and how increasingly attention is shifting to reducing vulnerabilities and increasing resilience.

The relationship between sustainable development and human development

The World Commission on Environment and Development (the "Brundtland Commission") defined sustainable development as progress that "meets the needs of the present without compromising the ability of future generations to meet their own needs."[2] This formulation was deliberate, to find a compromise between "post-Stockholm environmentalists" and those who believed that growth and development were priorities and that the demands of environmentalists served as barriers to those ends.[3] The definition avoided both the words "development" and "environment" and instead focused attention on three aspects: needs, the ability to meet the needs, and the link between present and future generations. Internal discussions within the commission recognized that sustainable development required that certain basic needs for all persons had to be met.[4]

Just as the dialogue on sustainable development focused on basic needs, the *HDR*s, from the very early years, also examined the relationship between human development and sustainable development. If human development has to ensure the expansion of the capabilities of *all* persons, then this "universalism of life claims" extends across time as well. Thus, "the strongest argument for protecting the environment is the ethical need to guarantee to future generations opportunities similar to the ones previous generations have enjoyed. This guarantee [was] the foundation of 'sustainable development.'"[5]

Moreover, from a human development perspective, preserving only natural capital would not be sufficient. As *HDR 1994* argued, "All postponed debts mortgage sustainability—whether economic debts, social debts or ecological debts."[6] Thus, for sustainable human development all types of capital—physical, human, and natural—had to be replenished to ensure their availability for the use of future generations.

Equity and responsibility

At the core of the formulation of sustainable development is the notion of intergenerational equity, to ensure that sufficient resources were available for future generations. This notion had a long history, and had been reflected in various international environmental instruments, such as the 1946 Whaling Convention, the 1972 World Heritage Convention, and, of course, Principle 1 of the Stockholm Declaration.[7] Proponents of human development wanted to make it clear that sustainable human development did not mean that the goal was to "sustain human deprivation." On the contrary, while resources had to be conserved for future generations, it could not be at the cost of giving less attention to the needs of the less privileged today. As a concept, sustainability focused on intergenerational equity but the "ethic of universalism clearly [demanded] *both* intragenerational equity and intergenerational equity."[8]

As with equity, the notion of responsibility is equally important in the cause of sustainable development. Multilateral environmental agreements have generally applied differential standards to different countries, sometimes by giving longer periods for compliance and sometimes by imposing different standards altogether.[9] The Montreal Protocol on the protection of the ozone layer gave a 10-year grace period to developing countries to comply with the standards, whereas the Kyoto Protocol under the UN Framework Convention on Climate Change (UNFCCC) did not impose any emission reduction obligations on developing countries at all. Such differentiation is born out of recognition of responsibility, that is the countries responsible for polluting must also be held accountable for the clean-up—if their actions had adverse consequences for others. The UNFCCC recognized that, while states had the sovereign right to exploit their own resources, they also had the responsibility "to ensure that activities within their jurisdiction or control do not cause damage to the environment of other States or of areas beyond the limits of national jurisdiction."[10] Thus, while intergenerational equity demanded that countries protect the climate system for the sake of future generations, intragenerational equity divided that common responsibility on a differentiated basis. This meant that developed countries "should take the lead" in combating climate change and its adverse effects.[11]

Consumption and sustainability

In fact, as intergenerational equity is linked to intragenerational equity, sustainable human development would imply that the world's income and consumption patterns also would have to be restructured.[12] At one end, it would be undesirable for poor countries to aspire to the unsustainable consumption patterns of the developed world. However, the burden of responsibility could not lie with poor countries alone, or even disproportionately. For sustainable development to be viable, rich countries would have to change their lifestyles as well.

Energy is an aspect of consumption closely associated with elements of human development as well as environmental sustainability. There is a strong correlation between income levels (GDP per capita) and electricity consumption. At the same time, higher energy consumption across the world is strongly based on fossil fuel use. In other words, while energy access is a basic human need, emissions of greenhouse gases largely have come from electricity generation, rising with higher development levels.[13]

Despite these close associations between concerns about sustainability and indicators of human development, conceptual gaps remain. At the heart of the two concepts is a fundamental tension. While human development is about expanding choices, it is not clear whether choices *per se* are fungible across time. Ricardo Fuentes-Nieva and Isabel Pereira argue that as choices are defined by current circumstances, the focus ought to be on the possibility of doing something today versus the possibility of doing something at a future date.[14] To that extent, it is not sufficient to improve living standards—it is also necessary to expand capabilities and entitlements on an equitable basis within and across generations.[15]

Trade, water, and energy as means; sustainability as one of the ends

In this chapter we examine trade, water, energy, and the environment as some of the means of human development. Sustainable development, in turn, is treated as one of the ends of human development, assuming that the capabilities and freedoms to exercise choices need to be expanded in the present and preserved for the future.

All four elements present policy-makers with tradeoffs between enabling human development today versus constricting opportunities and choices in future. Trade, as a means to increasing access to a wider set of goods, services, and ideas, could increase living standards and promote human development. But the rules of the global or regional trading systems, if iniquitous, could also lock out large numbers of poor people or poor countries from the benefits of trade, thereby making the expansion of choices in future even harder. Likewise, energy access is a means to securing education, health, and other essential services. But fossil-fuel-dependent energy militates against the development choices of poorer countries today and of future generations as well, by leaving little "carbon space" for either. Access to water is a central human development priority. But the management of water also demands paying attention to equity in access within communities and across borders. Finally, human beings are part of the natural environment, drawing on environmental services such as fishing resources, soil for agriculture, forests as carbon sinks, and so forth. The challenge is to account for these services as well as the negative externalities that human activities might impose on other human communities and other species.

Trade: with a human face

Trade impacts human development in several ways.[16] It can generate welfare gains by increasing the efficiency of resource allocation and making a wider range of products and services available at lower cost.

But overall income growth need not necessarily translate to overall human development progress if the opportunities and benefits of trade are restricted or if the costs associated with trade and economic restructuring are spread across the wider society. So, the distribution of the gains from trade matters.

In addition, trade can contribute to human development by increasing employment and broadening opportunities and capabilities. If trade-led growth also results in higher government revenues, which are spent on education or health care, then human development outcomes improve. The impacts on gender relations or on the environment are more ambiguous, depending on the trade patterns, the rights embodied for all parts of the labor force, or the provisions included to maintain environmental sustainability without undermining trade opportunities. Finally, not only is trade a route to rising incomes, but human development in the form of healthier, more educated, and more skilled people opens up new opportunities to benefit from trade with production and export of higher value-added products and services.

In other words, the relationship between trade and human development is mixed and contingent on specific conditions. As regards international institutions, the challenge over the past two decades has been to develop a trading system that allows countries and people to advance human development. Compared with environmental issues, trade and growth have served as a common thread in most global *HDR*s. Since the earliest *HDR*s, strong arguments have been made that, for example, the developing world has lost out far more from high tariffs on its agricultural and textile products compared with foreign aid transfers.

The international trade regime was ostensibly built on the principles of promoting the development of less advanced countries. Free trade is only a means to that end. The trade regime's history shows that its rules are not solely driven by the objective of free trade. The first sentence of the Charter of the International Trade Organization (ITO)—framed in Havana in 1947—pledged that members would work to attain "higher standards of living, full employment and conditions of economic and social progress and development."[17] The call for promoting economic development of countries in the early stages of industrial development, in Article 1.2, preceded that of achieving a reduction of tariffs and other trade barriers (Article 1.4).[18]

After the ITO's premature demise, the General Agreement on Tariffs and Trade (GATT) and the World Trade Organization had fewer formal linkages with broader development objectives, but some were significant nevertheless. The GATT recognized that the management of international trade had variable impact on development, emphasizing the concerns of less developed countries, and allowing developing countries to maintain flexible tariff structures.[19] The Agreement Establishing the World Trade Organization offered less flexibility and policy space, yet it cited the need to ensure that poor countries secured "a share in the growth in international trade commensurate with the needs of their economic development." The least developed countries would undertake commitments "consistent with their individual development, financial and trade needs."[20] In other words, how rules were interpreted and implemented depended on a member's stage of development and the competing objectives of social and economic development that it faced.

Human development: present at the creation?

The participation of developing countries in the first six rounds of GATT negotiations had limited substantive implications. Their main gain had been the Generalized System of Preferences (GSP), which exempted developing countries from the "most favored nation" (MFN) clause. But it hurt developing countries in the long run by encouraging their import-substituting strategies and allowing developed countries to impose GATT-inconsistent barriers to textile exports from developing countries. Worse still, being temporarily satisfied with these "special and differential" concessions, developing countries missed out on the chance to participate more substantively in negotiations.

The Uruguay Round of trade negotiations, which culminated in the creation of the World Trade Organization, was the first time that developing and developed countries negotiated on a set of trade rules and procedures that would apply to all countries, under a "single undertaking." Although the conceptual frameworks for human development were being developed around the same time, there was little direct application of the concepts to international trade debates. However, the structure of the trading system soon opened up a plethora of issues that went to the heart of concerns raised by the human development approach.

Trade as enabler; trade as the disabler

The argument that trade could be the pathway out of poverty gained salience with the economic success of East Asia: first Japan in the post-Second World War period, followed by South Korea, Taiwan, Hong Kong, Singapore, and eventually by Thailand, Indonesia, and

Malaysia. Perhaps the strongest evidence came from China, which, having followed insular and disastrous economic policies in the late 1950s and 1960s, began a process of economic reform in 1979. The result is that it is now the world's largest trader and has lifted more people out of poverty than any other country in history. The patterns of growth varied across the region—greater focus on supporting large domestic conglomerates in Japan and Korea, encouraging investment by multinational firms in Singapore and Malaysia, or growth driven by state-owned enterprises in China. But all of these economies had common starting points: investments in physical and human capital, declining fertility rates allowing women to enter the formal workforce in large numbers, and increasing agricultural productivity allowing more labor to come off the farms and enter the industrial sector.[21]

However, there was fierce disagreement about other factors, particularly the role of the state in pushing such growth. The prevailing "Washington Consensus," driven by the World Bank and the International Monetary Fund, was that the East Asian Miracle was the result of high savings and investment rates and open economies. Critics challenged this reductionist theorizing, and suggested that the international financial institutions had chosen not to study the region in depth and that the "miracle" economies grew not because of a minimalist role of government but thanks to active state intervention.[22] When the East Asian financial crisis hit in 1997–98, these opposing views blamed each other for worsening the economies. One set blamed the crisis on crony capitalism while critics of the international financial institutions argued that the structural adjustment policies imposed on these economies had made matters worse, by hacking government spending and reducing investment exactly when it was needed.

The experience of the East Asian countries had an impact on how trade was viewed as well. The crisis occurred only two years after the WTO was born. In celebrating the economic growth of the region, many had overlooked that these economies were aggressive in promoting exports but were far more reticent in opening their markets to imports. This form of free trade, for the Asian tigers, was a mercantilist means of promoting growth, not a dogma to be followed blindly.

Others argued that it was not just the East Asian economies but also the industrialized countries which had followed restrictive trade policies during similar stages of development. For instance, between 1816 and 1945, the United States had some of the highest tariff rates on imports of manufactured goods in the world.[23] Similarly, developed countries did not have all the institutional arrangements—universal suffrage, professional bureaucracies, property rights, corporate governance and competition commissions, etc.—during their development periods, which were being cited as necessary for growth in the developing countries.

The WTO's early stages were characterized by tensions around these intellectual positions, not only the call for reduced tariff and non-tariff barriers but the demand for more institutional changes, such as the apparatus for intellectual property protection. Despite the hard-fought Uruguay Round of negotiations, developing countries viewed the WTO as intervening in their domestic institutional structures about the same time as the aftermath of the East Asian crisis was calling into question the merits of the Washington Consensus. The introduction of new trade issues was vociferously opposed and rejected. At the Singapore Ministerial Conference (9–13 December 1996), India and other developing countries blocked negotiations on trade and competition, trade and investment, government procurement, and trade facilitation and argued that the International Labour Organization was competent to deal with labor standards. Meanwhile, nongovernmental organizations in developed countries also were questioning the legitimacy of the WTO, with allegations of weak protections for labor standards and the environment. Things came to a head when thousands of anti-globalization protesters overshadowed the ministerial meeting in Seattle in December 1999. The Seattle Ministerial failed because of disagreements between the EU and the United States. But developing countries also strongly protested against the inequities in the WTO. The 1990s, which had begun with hope for a peace dividend at the end of the Cold War, ended with deep disagreement on the patterns and terms of trade and deep suspicion about the alleged benefits for human development.

Paradigm shift

It was against this background that *HDR 1999* was published, titled "Globalization with a Human Face." It argued that globalization was offering new opportunities for countries to attract investment and technologies and access new markets. However, despite deeper integration with the world economy (measured, say, by share of trade as a proportion of GDP), many countries remained vulnerable to the "vagaries of global markets," seeing little rise in incomes (80 countries had registered a fall in per capita incomes during the 1990s), few job opportunities, rising inter-regional disparity within countries, and a growing gap between the world's richest and poorest citizens.[24] This kind of lopsided globalization was also responsible, the *HDR* argued, for new threats to human security (financial volatility, job insecurity, health insecurity, and even communal and social tensions).

In response, at the core of a new kind of globalization was the need to reinvent national and global governance. The elements of such a reinvention included making human well-being the end, with open markets and economic growth as the means.[25] This meant that policies were needed to respond to changing labor markets and shrinking government fiscal resources. They were also needed to reduce financial volatility by incrementally opening up financial markets and imposing strong supervision of the banking sector. Moreover, technological development, while essential, had to be nudged towards innovations that would promote human development and reduce poverty. Intellectual property rules had to be modified if they conflicted with the human development aims of improving health outcomes or making knowledge accessible via new technologies to hitherto unconnected countries and people.

This new kind of globalization would also be more inclusive in structures of global governance, giving poorer countries more of a voice in setting agendas and determining outcomes. It is hard to imagine, given that we now live in a world of the G20 and with China or India or South Africa at the top table of almost any international negotiation, how exclusionary global governance processes were less than two decades ago. Despite shifts in economic trends and governance arrangements in the 2000s, deep democratic deficits still remain. But the *HDR* was at the forefront advocating a paradigm shift.

Agenda-setting

The reshaping of the agenda for a more human-centered international trade system can be witnessed in three areas: intellectual property and access to medicines; the treatment of indigenous knowledge; and building capacity for trade.

TRIPS and public health

Intellectual property rights (IPRs) were only indirectly dealt with under the GATT. There was no provision in the GATT that mandated the protection of IPRs. Instead, Article 20(d) permitted members to make exceptions to the application of GATT rules with the objective of protecting "patents, trademarks and copyrights, and the prevention of deceptive practices."[26]

Eventually, developments within the United States were instrumental in bringing IPRs onto the GATT agenda. Private sector lobbying for improved protection of intellectual property had been a feature of the Tokyo Round itself, with the Anti-Counterfeiting Coalition at the forefront, which then widened its scope of demands in the early 1980s. The pressure from domestic constituencies affected American trade policy. The US Trade and Tariff Act 1984 strengthened Section 301 of the Trade Act 1974, giving the US president the authority to impose sanctions on countries that did not provide "adequate and effective protection" to IPRs. The Intellectual Property Committee (IPC), representing research-based American industries, intensified lobbying efforts from March 1986. Consequently, IPR and trade negotiations were linked. The United States demanded stronger IPR protection through the GATT in return for access to developed country markets.[27]

In the early stages of the Uruguay Round, the United States and the European Community had specific demands on intellectual property: substantive standards; border and internal enforcement measures; application of GATT principles of national treatment, non-discrimination, and transparency; and an effective dispute settlement mechanism.[28] In turn, India and many other developing countries had a hardline position against these initiatives, arguing that IPRs were outside the mandate of the negotiations and wanting other international organizations, such as the World Intellectual Property Organization (WIPO), the UN Educational, Scientific and Cultural Organization (UNESCO), and the UN Conference on Trade and Development (UNCTAD) to deal with them. Secondly, they did not want enforcing IPRs to become a barrier to legitimate trade.[29]

At the heart of these positions was the interest in keeping prices of medicines low, in addition to promoting domestic industry. At a World Health Organization (WHO) conference in Geneva in May 1981, Prime Minister Indira Gandhi had declared: "My idea of a better ordered world is one in which medical discoveries would be free of patents and there would be *no profiteering from life or death.*"[30]

In 1998, the Pharmaceutical Manufacturers Association of South Africa, together with 39 pharmaceutical companies, initiated a three-year legal battle to prevent the South African government from implementing its 1997 Medicines and Related Substances Act. The law permitted parallel imports of drugs from cheaper markets and the grant of compulsory licenses to drug producers in other countries. Antiretroviral therapy using patented drugs for HIV-positive persons was prohibitively costly. At $10,000 per person per year, it would have cost the South African government 27 times its entire budget on medicines to treat less than a sixth of its HIV-positive population.[31] An Indian company, Cipla, offered the same package of drugs for $350. A coordinated campaign by civil society groups such as the Treatment Action Campaign, Oxfam, and Médecins Sans Frontières (MSF), and pressure from the European Parliament forced the pharmaceutical companies to announce in April 2001 that the law was TRIPS-consistent.[32]

In another case, the United States alleged on 30 May 2000 that Brazil's industrial property law of 1996 (Law No. 9.279) was inconsistent with TRIPS because it demanded the "local working" of a patent.[33] If the patent was not worked locally the law authorized the government to grant compulsory licenses. Brazil had used the law to acquire cheaper generic versions of antiretroviral drugs. Consequently, AIDS-related deaths had been halved since 1995. In June 2001 the United States agreed to withdraw the case while Brazil promised consultations before enforcing any contentious laws.

These cases affected the thinking in international organizations on access to drugs in poor countries. Resolutions demanding equal access to drugs through parallel imports, licensing, and domestic production were passed by the UN Human Rights Commission, the World Health Assembly, and the UN General Assembly, in April, May, and June of 2001, respectively. These developments confirmed that IPR protection was not the exclusive domain of the WTO or WIPO.[34]

Taking up the cue, the African Group, Brazil, India, and 15 other countries made a radical statement that "nothing in [TRIPS] reduces the range of options available to Governments to promote and protect public health."[35] India argued that affordable access to medicines was "a fundamental human right."[36]

In the run-up to the Doha Ministerial Conference of 9–14 November 2001, the United States and the European Union were keen to launch a new trade round. The Seattle Ministerial in December 1999 had failed because of disagreements between them. But developing countries had also strongly protested against the inequities in the WTO. The terrorist attacks on 11 September 2001 made it imperative that a debacle like Seattle not recur. Developing countries now had the opportunity to extract greater concessions from the United States.[37] The threat of anthrax attacks gave greater credibility to their case, as even developed countries could face health emergencies.[38] Eventually, a compromise Declaration on the TRIPS Agreement and Public Health was drafted by eight countries at Doha.[39]

Although the Doha Declaration secured a consensus, it postponed any deal on compulsory licensing in the case of countries without sufficient manufacturing capacity for drugs. Developing countries did not gain any significant concessions in comparison with their original demands. The only major concessions were on the issue of "exhaustion" of IPRs and the 10-year extended transition period for LDCs.[40] On compulsory licenses, although countries retained their rights the legal dimension was not very flexible. Countries still had to prove that the remedial measures against health emergencies constituted the "least possible interference" with IPRs.[41]

On 30 August 2003 the WTO's 146-member General Council took a decision to ensure access to cheap drugs for countries without a sufficient pharmaceuticals manufacturing capacity. Brazil, India, Kenya, South Africa, and the United States negotiated the deal. The "decision" temporarily waived certain TRIPS obligations but demanded proof of the "eligible" importing member state's insufficient capacity to produce the medicines and limited the exporting member's supply only to the amount necessary to meet the emergency needs.[42] Although several NGOs criticized the deal as "a gift bound tightly in red tape" designed to delay emergency drugs delivery,[43] the declaration was still a breakthrough and set a precedent for how a human development lens could force the renegotiation or reinterpretation of international law.

Indigenous knowledge

Another human development dimension was the preservation of traditional knowledge through community rights. Although human development is the expansion of individual choices, *HDR 2004* argued that community-based rights need not necessarily conflict with the expansion of freedoms. On the contrary, the suppression or non-recognition of community rights could, at times, undermine the individual's multiple identities or deny the person access to resources that could benefit them even on an individual basis.

The idea of multiculturalism in globalization was premised on four principles:[44]

1 Recognizing that blindly defending tradition could hold back human development.
2 But respect for difference and diversity was essential.
3 Diversity recognizes people's multiple identities and, in fact, thrives in an interconnected and interdependent world.
4 Addressing imbalances in political power could help poorer or weaker communities to protect their cultures and heritage.

These principles were put to the test with regard to the protection of traditional knowledge and the application of international trade rules.

In March 1995, the US Patent and Trademark Office (USPTO) granted a patent (no. 5401504) to two scientists from the University of Mississippi for discovering the wound-healing properties of turmeric. However, the medicinal benefits of turmeric had been known in India for centuries. It was not only used as a home remedy for wounds and rashes but also was used in Indian traditional medicine systems such as *ayurveda*. Patents had been granted for other Indian herbs and plants as well, whose medicinal uses were commonly known. Some of the more prominent examples were *jar amla* (*Phyllanthus niruri*) for hepatitis, *karela* (bitter gourd) for diabetes, and *neem* (*Azadirachta indica*) for pesticides. However, these had not been challenged.

The Indian Council for Scientific and Industrial Research (CSIR) decided to appeal the turmeric patent on the grounds of "prior art"—that is existing public knowledge. Apart from other citations, seventeenth-century Persian and ancient Sanskrit texts were presented to argue that the healing powers of turmeric were common knowledge in India. Failing to satisfy the novelty and inventiveness criteria, the turmeric patent was revoked. This was the first known case where the patenting of traditional knowledge had been successfully challenged in the United States. It created awareness within India that its common resources and knowledge were the community's intellectual property and efforts to prevent their appropriation could be successful.

Yet, other instances continued to emerge around the world. By one estimate (in March 2000), 7,000 patents had been granted for unauthorized use of traditional knowledge. These included the medicinal properties of the sacred Ayahuasca plant in the Amazon basin; the Maca plant in Peru, which enhances fertility; and a pesticidal extract from the neem tree used in India for its antiseptic properties.

HDR 2004 shone a light on how deep integration within the global economy was creating a sense of siege for many indigenous communities, especially in the context of extractive industries and traditional knowledge. Intensive resource extractive industries operating in areas where indigenous communities are concentrated often ignore the cultural significance of the lands to the communities. San Bushmen in Botswana have opposed exploration licenses in the Kalahari; more recently, bauxite mining has been opposed in the Niyamgiri hills of Odisha, India. Beyond the cultural or religious connections to land, indigenous communities also have opposed extractive industries because of exclusion from decision-making processes, concerns about adverse economic and environmental impact on the community, and limited or broken promises about compensation for appropriation of natural resources.

HDR 2004 also argued that intellectual property regimes failed to recognize collective ownership of knowledge, allowing others to use traditional knowledge, apply for patents by claiming that they had developed novel products, and benefit from commercial gains with none accruing to the community.[45] The issue was further complicated by contradictory rules within different international institutions. The Convention on Biological Diversity recognized traditional knowledge by demanding that contracting parties preserve and maintain the innovations of local communities and that wider application of such knowledge be based on their approval and involvement. By contrast, the World Intellectual Property Organization and the TRIPS Agreement under the WTO did not afford such rights.

The human development approach took a middle path, by arguing that the solution did not lie in blocking flows of investment and knowledge. Instead, there was the need to create equitable socio-economic opportunities within a democratic framework, which protected liberties (including community rights). In this vein, the first step was to recognize indigenous people's rights over their resources. *HDR 2004* highlighted national legislation in the Philippines requiring consent for access to ancestral lands, and Guatemalan law permitting use of traditional knowledge by placing them under state protection. The second step was to require participation and consultation with local communities, which meant providing complete information and disclosure (such as where plants originated before granting patents based on them). The *HDR* also called for documenting traditional knowledge without prejudicing the rights of the community. Thirdly, benefits had to be shared with the communities, whether in mining projects or in using intellectual property rules in innovative ways, such as industrial designs to protect the Kazakh carpet industry, geographical indications for liquor and tea in Venezuela or Vietnam, or copyrights and trademarks for indigenous art in Australia and Canada.

Although conflicts and tensions persist in how the natural and intellectual resources of indigenous communities are exploited or expropriated, viewing these resources as means to expand choices and opportunities for both the community and the global economy at large was one way to mitigate risks of exploitation and increase joint benefits.

Capacity for trade

The cases of intellectual property and public health or that of indigenous knowledge being valued or not, demonstrated that developing countries often struggled with the capacity to gainfully engage in the international trade system. *HDR 2005* railed against the iniquities in global trade, in terms of barriers to market access for the least developed countries, erosion of preferential trading arrangements, highly skewed subsidies in agriculture benefiting farmers in developed countries, and increasingly restricted space for national governments to determine industrial policy, intellectual property rules or for trade in services.[46]

But greater market access or more equitable rules would not translate automatically into increased exports or human development gains, unless the capacity was there to take advantage of market opening. The *HDR* criticized the way technical assistance for capacity building was structured: pushing donor priorities, offering biased advice, miniscule funding, and little linkage to overall development policy.[47]

On aid, trade, and conflict in an unequal world, *HDR 2005* was published with the 2005 UN World Summit in mind. The summit had many purposes: a celebration of 60 years of the UN, a review of progress of five years of efforts towards the MDGs, and introspection on whether the existing architecture and institutions of global governance were indeed fit for purpose. The summit's outcome document reflected a wide range of concerns across a host of development and security issues. References to trade in the document made it a point to enable fuller participation by least developed countries and building their capacities. However, no concrete proposals were made.[48]

Nevertheless, the impact was beginning to be felt within the World Trade Organization. Within three months of the World Summit, in December 2005, the WTO's Hong Kong Ministerial established an Aid for Trade (AFT) initiative.[49] The Ministerial Declaration asked the WTO director-general to establish an AFT task force that would recommend how this new form of aid could be operationalized. Within a year, several developed countries substantially increased their commitments for aid-for-trade.[50] Furthermore, to ensure that commitments were indeed kept, a robust monitoring mechanism was designed: the OECD's Development Assistance Committee (DAC) would assess global flows; donors would submit self-evaluations; and recipients would prepare in-country assessments. The WTO would have overall charge of AFT surveillance, an extension of its monitoring responsibilities.[51]

Now about $25–30 billion of annual commitments for overseas development assistance (as calculated by the OECD) are directed for trade-related activities.[52] Although concerns have remained on whether AFT has had the desired impact, there is no doubt that aid directed specifically for improving the capacity of developing countries is now well-established as a priority of the WTO in general, and the Doha Development Round of trade negotiations in particular.[53]

Water: for life, for livelihoods, and for all

Water security is an integral component of human security. Access to clean and affordable water and sanitation is necessary for a healthy life and also for human dignity. Water is also a productive resource, shared within countries and across borders, and a driver of incomes and livelihoods directly for billions of farmers, and, indirectly, for industry and commercial services. *HDR*s have drawn attention to both dimensions but have extended the discourse beyond the physical availability of water to the social and political inequities, which result in vast inequalities in access to water for life or livelihoods.

Paradigm shift: reframing access

HDR 1998 highlighted deep gaps in access to water and sanitation. In rapidly developing countries, there was also the challenge of declining water quality (from both human and industrial waste).[54] The report drew attention to the links between water and other environmental resources, including land, forests, and pasture. It also explained that the links to poverty and environmental degradation went beyond income and were affected by rights of ownership of resources, institutions, ecological risks, and gender (particularly the time spent in collecting water and other resources).[55] Moreover, the report highlighted the potential stresses on water as a result of climate change and the impact on agricultural output and food security, especially in equatorial regions.

Every *HDR* has highlighted deep national and global deficits in access to water. Deprivations in access to improved water sources were already included in the Human Poverty Index for developing countries (HPI-1). In the year 2000, the Millennium Development Goals (MDGs) were announced, with one of the targets being halving the proportion of people without sustainable access to safe drinking water by 2015. In fact, the water target (and later that on access to sanitation) was included under the goal for ensuring environmental sustainability, recognizing the links between the human and the physical environment, especially with regards to a critical resource like water. The MDGs gave the *HDR*s an opportunity to highlight these deficiencies against the backdrop of global goals.

The emphasis was, however, less on the specific target and more on the need to draw attention to a basic deprivation. As noted by *HDR 2003*, which focused on the MDGs, "Whether the numerical target of a global goal was achieved is an important but inadequate measure of success, because it does not indicate whether setting the goal made a difference. In many cases enormous progress has been made [such as in increasing access to water] even though numerical targets have not been reached."[56] The report emphasized that rather than seek to only increase access, the resource had to be managed more holistically, drawing on appropriate and cost-effective technologies, making provisions for operations and maintenance, limiting environmental damage, and increasing equity. It paid particular attention to how urban consumers could be charged for the full cost recovery (capital costs and operations), so that resources could be generated to pay for increasing access to those with limited ability to pay. It also suggested that women could be trained in building and maintaining installations, such as hand pumps or toilets.[57]

Despite the emphasis on basic deprivation (or progress made), water remained out of the ambit of regular political discourse. *HDR 2006* sought to change the terms of the discourse by placing poverty, inequality, and power politics at the center of the "silent crisis" of water.[58] First, the 2006 report drew attention to the fact that wealth mattered in access to water, even as sanitation lagged further behind. As countries got richer, access to water did tend to increase—but this was not always the case. In several countries, higher relative incomes did not automatically translate into improved access to water or sanitation. Whether it was China in comparison with Vietnam (in the case of water) or India in comparison with Bangladesh (in the case of sanitation), national income was only a partial predictor of access. One of the problems was that national data systematically under-reported lack of access, either because many citizens lived in informal settlements not counted within the coverage areas of water utilities, or because the quality of the infrastructure and poor maintenance was seldom assessed.[59] It was not merely poverty that was the barrier. Deep-rooted inequalities in societies were responsible for many to remain outside the scope of the provision of basic services.

Secondly, the report calculated the human development impacts of the water and sanitation deficit. For those who argue that the focus of national policy ought to be increasing national incomes *before* other social development objectives are pursued, the analysis found that developing countries' GDP was 2.6 percent lower thanks to the water and sanitation deficit.[60] Furthermore, not only were these failures adversely impacting other MDGs, such as reducing infant and child mortality (diarrhea accounted for 1.8 million child deaths annually), but repeated bouts of illness would diminish lifecycle opportunities for the poor: children dropping out of school early; stunting; and cognitive deficiencies. All of these disabilities would translate into lower incomes.[61]

Thirdly, the disproportionate burden on the poor was a major focus of the report. It found that half of the 1.1 billion people without access to water were in the bottom 40 percent of the global income distribution. Using household survey data, the *HDR* reported that 660 million people had very limited capacity to pay for water connections. In sanitation, too, the poorest 40 percent of households accounted for half the share of the global sanitation deficit of 2 billion. In country after country, an individual's position on the income pyramid determined whether access to water was through pipes or from unprotected wells or informal tankers, or whether sanitation meant a flush toilet or pit latrines or no facilities at all.[62]

Yet, the poor paid more than the rich for rudimentary access to water and sanitation. The *HDR* busted the myth that the poor were unwilling to pay for water and sanitation services. It equally exposed the argument that mere willingness to pay would ensure that such basic services would be automatically provided. It argued that willingness to pay was different from ability to pay, especially when spending on water accounted for a large share of the household budget. And yet, slum dwellers across the world were paying 8–16 times more per unit of water to informal vendors and tankers compared with what richer households paid to water utilities.[63] Such regressive pricing also meant that richer households were able to consume a lot more water than poorer ones, further exacerbating inequalities in access (not merely to connections but the actual quantum of water used). Inequalities between rural and urban areas, between men and women (or boys and girls), and between social classes and castes also were highlighted in the report.

In other words, *HDR 2006* acknowledged the global deficit in water and sanitation but recast access as a function of not merely the availability of water but one that was disproportionately hurting the poor, forcing them to spend more (in absolute and relative terms), impacting household and national incomes, undermining health outcomes, and exposing the argument that income growth would be sufficient to achieve other human development outcomes.

Agenda-setting

The human development approach guided thinking and practice on water and sanitation in at least two ways: the International Decade for Action "Water for Life" (2005–15); and how tensions over transboundary waters were viewed.

International Decade for Action "Water for Life" 2005–15

In recognition of the Millennium Summit in 2000, from which the MDGs emerged, and the 2002 Johannesburg World Summit on Sustainable Development, the UN General Assembly agreed in December 2003 that an International Decade for Action "Water for Life" would be observed during 2005–15. The goal was to "promote efforts to fulfil international commitments made on water and water-related issues by 2015" with a focus on long-term sustainable management of water resources.[64] UN-Water, which was in charge of coordinating the "Water for Life" decade, launched two initiatives on capacity building and a program on advocacy and communication.

By focusing on water in the first year of the decade, the *HDR 2006* took the intellectual leadership in framing the issues, which could be discussed and acted on, by the international community, national governments, and individual citizens. The first report from the UN secretary-general, containing pledges of activities from various UN agencies, did not contain any reference to inequality.[65] Even references to sanitation were merely repetitions of the MDGs. But by December 2006, the General Assembly had drawn special attention to sanitation by naming 2008 as the International Year of Sanitation (IYS). The declaration stated that it was important to confront sanitation issues "in complementarity with water."[66] IYS 2008 was primarily for advocacy, with stated aims of increasing awareness, mobilizing national and local governments, increasing funding allocations, and building capacity.[67] By the time the mid-term review of the International Decade occurred in 2010, member states were calling for "Sustainable Sanitation—the Five Year Drive to 2015."[68]

Another key achievement was the declaration of water as a human right. *HDR 2006* demanded that recognizing water as a human right should be the first priority for international and national policy. It argued that a rights-based approach would follow the principles of "equality, universality and freedom from discrimination." It would also ensure that income level, ability to pay, place of residence, or membership of a social group would not be the basis for excluding anyone from this basic service.[69] Recognizing the right to water would provide the poor with the moral claim to demand, through social, political, and legal channels, their entitlement to a fair share of the resource.

On 28 July 2010, the UNGA, through Resolution 64/292 recognized access to safe and clean drinking water and sanitation as a human right.[70] In 2002 the UN Committee on Economic, Social and Cultural Rights had adopted a General Comment on the human right to water, but this was a non-binding, normative framework. *HDR 2006* had recognized that reframing this framework as a substantial public policy priority would be a challenge. By 2008, the UN Human Rights Council had adopted a resolution to appoint an independent expert on the "human rights obligations related to access to safe drinking water and sanitation."[71] The demands of achieving the MDG targets on water and sanitation, the increasing focus on exclusion and inequality in access, and the need for some tangible results during the International Decade resulted in the formal recognition of the right to water and sanitation. The momentum has continued to build across the UN system with further resolutions from the World Health Organization (in May 2011[72]) and again from the Human Rights Council (in September 2011[73]), and reports from the special rapporteur on the right to safe drinking water and sanitation.[74]

Transboundary waters and human development

Another key contribution of the human development approach was with regards to transboundary waters. It is well known that international water basins—catchments, lakes, shallow groundwater—cover more than half the Earth's land surface. As a result, 40 percent of the world's population lives in these transboundary basins. There are 263 such basins globally, affecting 145 countries. One consequence of this deep hydrological interdependence is the belief in some quarters that wars over water resources would be the major cause for conflict in the twenty-first century. Others have pointed out that water wars have not occurred for thousands of years.

HDR 2006 argued that the obsession with proving or disproving the water wars hypothesis was a distraction. Instead, the focus had to be on human security rather than narrowly defined national security concerns. It argued that cooperation or conflict over water across international borders were choices, which would be determined by how societies recognized their hydrological interdependence, their shared vulnerabilities, the threats of ecological disasters and their ability to find win-win solutions in managing shared water resources.[75]

A human development lens on hydrological interdependence would start, first, by tracing the nature of economic activity across the entire transboundary basin. The report demonstrated this through cases of rivers such as the Nile and the Mekong and of lakes such as Lake Victoria in east Africa, Lake Chad in west Africa, and Lake Titicaca in South America. For instance, the Mekong River is the source of hydropower in its upper reaches for China and Myanmar—but more than 40 percent of the hydropower potential in the lower Mekong lies in Laos. At the same time, half of Thailand's arable land lies within the river basin, which also accounts for half of Vietnam's rice production. A significant share of the Cambodian population, in turn, depends on the Tonle Sap Lake.[76]

Secondly, the report documented the human development consequences of not cooperating among riparian states. This was particularly challenging when economic activities were similar in nature on either side of the border. One-fifth of Turkey's irrigable land lay within the region where the Tigris and Euphrates rivers originated. But a fifth of Syrians also lived within the Euphrates catchment area, and further downstream major Iraqi cities were on the banks of both rivers. Diversions of water in one country (say, in China for agricultural purposes) were considered threats to national interests elsewhere (such as in Kazakhstan). Such tensions occurred within OECD countries as well, such as in the Colorado and Rio Grande basins shared by Mexico and the United States.[77] The consequences of non-cooperation included severe ecological disasters (as has happened in the Aral Sea in Central Asia, or Lake Chad in Africa), fall in agricultural yields in Central Asia, adverse health impacts, and the export of pollution across borders.

Thirdly, *HDR 2006* overlay the human development consequences of unresolved political disputes, which affected access to water. This was particularly demonstrated in the case of Israel and the Occupied Palestinian Territories (now the State of Palestine). The report highlighted deep inequalities between Palestinians, Israeli citizens, and Israeli settlers in the West Bank and Gaza Strip. per capita differences in water availability ranged from four to nine times. This was a result of the absence of defined rights to access the Jordan River as well as extraction of water from aquifers before they flowed into Palestinian-controlled areas. These inequalities impacted availability of water for livelihoods, particularly agriculture, in addition to education and health outcomes.[78]

Yet, *HDR 2006* also made the case for cooperation, even in the most politically disputed regions. It found that well-established rules, combined with strong capacity and institutions could offer numerous benefits. Although many water disputes occurred over the quality of water, cooperative events included water quantity but also infrastructure, quality, hydropower, joint management, flood control, etc.[79] Ultimately, the conditions for cooperation (there are more than 295 international water agreements) depended on jointly assessing human development needs, building trust and legitimacy through joint projects, and adequate financing to build credible institutions to manage and monitor shared water resources.[80]

In 1997 the UN codified the principles of cooperation under the UN Convention for the Non-Navigational Use of Shared Watercourses, which were themselves built on the 1966 Helsinki Rules. Following the publication of *HDR 2006*, transboundary waters continued to get attention during the International Decade for Action "Water for Life." In March 2010, the UN Secretariat proposed that cooperation over water had benefits for human security. It drew on the framing of the *HDR* of direct benefits from the rivers, lakes, and aquifers, benefits to the water bodies, the potential for easing broader political tensions, and serving as catalysts for greater economic integration between states.[81] Subsequently, the UN declared 2013 as the International Year of Water Cooperation.

More importantly, all the core principles of strengthening cooperation—joint monitoring and data sharing, building trust, investing in institutional and human capacity, enforceable legal instruments, raising more financing, and sharing the benefits of basin-wide management—have now become embedded in how transboundary water issues are discussed at the international level. These principles might not have been translated into resolutions of all cross-border water-related disputes, but the shift from a focus on water wars to attention towards human security, inequality, and joint benefits was possible because of an intellectual approach that prioritized individuals and their entitlements.

Energy and environment: access not excess

The tension between development and the sustainability of the natural environment has been long running in national and international policy debates. When the UN decided to convene a conference to commemorate the 1972 Stockholm Conference, developing countries had demanded that it be called the UN Conference on Environment *and* Development; subsequently other major conferences also have tried to articulate this balance.[82] As this chapter argued earlier, for human development to ensure the expansion of capabilities across time, the natural environment had to be conserved for use by future generations. Equally, for sustainable development to have true resonance across human communities, there had to be sustainability of physical and human in addition to natural capital. Sustainable development could not mean locking people into lives of poverty. At the international level, reconciling these tensions required emphasis on equity, responsibility for past behavior (pollution, greenhouse gas emissions, etc.) and a consciousness of current and future behavior (by changing patterns of consumption).

The relevance of a human-centered paradigm for sustainable development was articulated well before the Brundtland Commission and the UN Conference on Environment and Development (the Rio Earth Summit) in 1992. In the lead-up to the 1972 Stockholm Conference, developing countries did not believe that it was their responsibility to resolve issues of environmental pollution and resource scarcity. The developed world bore sole responsibility. Mahbub ul Haq suggested that a concept of environmentally sound, people-centered development could be proposed. This was the foundation of the Founex Report on Development and Environment, which Haq authored. The report helped to bridge the divide between developed and developing countries for the purposes of the Stockholm Conference and Haq's formulation became a forerunner to the concept of sustainable development.[83]

Early *HDR*s struggled to incorporate considerations of the physical environment in their quantitative analyses. The first report, *HDR 1990*, acknowledged that there were conceptual and methodological challenges in quantitatively measuring issues such as political freedom, personal security, and the physical environment. Yet, it urged that analyses of human development ought not to ignore such issues, which might add to the "qualitative dimensions of human life." It also pointed out that there was a close relationship between poverty and environmental degradation, which in turn could be a cause for reversing gains in human development. The report emphasized that rich countries had far higher levels of overall industrial pollution but poorer ones were gradually accounting for the greatest increase in new pollution and did not have the resources to adopt more benign technologies.[84]

Global environmental concerns were populating the international agenda in the 1990s. The Montreal Protocol on Substances that Deplete the Ozone Layer was signed in 1987 and entered into force in 1989. It was, however, the 1990 London Amendments to the protocol that promised a Multilateral Fund to developing countries (Article 5 countries in the protocol) to pay for the incremental costs of shifting away from ozone-depleting chlorofluorocarbons to alternative chemicals. In 1992 the Rio Earth Summit resulted in several new conventions: on biological diversity, to combat desertification, and the most prominent one on climate change. That year, *HDR 1992* lamented that while industrialized countries were worried about the ozone layer and global warming, the core issue was excessive consumption of natural resources. For poorer countries, the worry was not about excess but scarcity, of good quality water or land.[85]

But seizing the international moment that was the early 1990s, the report also articulated the human development vision for building global institutions fit for the twenty-first century. Among them were policies for sustainable development. It was quick to critically analyze faults within the newly created Global Environment Facility (GEF), calling for a broadening of its mandate to deal with issues of concern to developing countries (desertification, urban degradation, water pollution), much greater funding (from $800 million to $5–10 billion), and alteration of the governance of the GEF to permit more participation by developing countries. Other ideas were promoted, such as greenhouse gas permits for all countries (the precursor of the Clean Development Mechanism under the UNFCCC), fossil fuel consumption taxes, and pollution taxes to raise the funds needed to support transfer of technology to poor countries.[86] Many of these ideas have continued to inform debates in environmental regimes. Although the quantitative indices were not capturing environmental externalities and costs explicitly, the human development approach, nevertheless, was seeking to reconcile development and the environment by putting people at the center.

Paradigm shift: environmental sustainability for the poor

A people-centered approach to sustainable human development offered ways not only to reconcile the priorities and negotiating positions of rich and poor countries, but it also made it possible to imagine new policy paradigms. These emerged in at least four ways: human security; the impact of environmental degradation on the poor; different ways to view the present and the future; and sustainable consumption.

Human security and the environment

If human development is about expanding people's choices, human security is the condition in which people feel they can exercise those choices freely and safely and that opportunities available to them today would not be completely lost in future.[87] Moreover, as an integrative concept, human security implied that one's security was intimately tied to solidarity with another's. The universalism of life claims would not accrue otherwise.

Both these interpretations—of feeling secure over time, and deriving security through solidarity—lie at the core of how environmental externalities are understood and confronted. Thus, environmental security was an integral part of human security. *HDR 1994* emphasized that local environmental concerns were no less important than those dominating international negotiations. It raised red flags over water scarcity (as a factor behind political tensions), deforestation on an industrial scale leading to desertification, salinization of irrigation systems, and air pollution as worsening particularly in developing countries. The gradual erosion in natural resources—particularly land and forests—made natural disasters even more likely and devastating because the resilience of societies, economies, and physical infrastructure was dependent on resources and capacity.

In addition to these local environmental challenges, at least three kinds of global environmental externalities also were evident by the early 1990s, namely air pollution, ozone depletion, and climate change. The concern was not merely environmental but also the disproportionate impact these externalities could have on regions that were not responsible for emitting the pollutants in the first place.[88]

Environmental degradation and the poor

International negotiations on environmental issues were (and still are) often caught in a bind between the North (asking poor countries to assume disproportionate responsibility) and the South (claiming that problems created by rich countries were theirs to solve). The human development approach, instead, focuses on how people are affected by environmental degradation. As was described in the case of transboundary waters, following the lives and livelihoods of communities would help to assess these impacts. As *HDR 1998* analyzed, "poor people and environmental damage [were] often caught in a downward spiral."[89] Forced to deplete resources to survive, the resulting environmental degradation traps many in further cycles of poverty. The desperation arises from poorly defined ownership rights over natural resources, ineffective institutions, and efforts to minimize risk and time devoted to dealing with basic resources.

HDR 1998 demonstrated how environmental impacts affected the poor disproportionately. It explored the poverty-environment nexus for water pollution, air pollution, domestic solid waste, industrial hazardous waste, soil degradation, deforestation, and loss of biodiversity. Water pollution, partly because of poor sewage treatment, was responsible for curtailing the life chances of millions of children (a point further explored in *HDR 2006*). But it was also affecting livelihoods, particularly of fishing communities in Asia and Africa. Similarly, air pollution is often considered a problem for industrialized countries. Yet, even in the mid-1990s, 90 percent of deaths caused by air pollution were occurring in developing countries, most of which were a result of indoor air quality in rural areas.[90] These debates have continued, sometimes unhelpfully, with emerging economies now trying to prove that their air quality is better than in others' while failing to recognize the scale of the problem.[91] Soil degradation also had direct impacts on the poor because nearly a third of the world's population depended on subsistence farming. The loss of forests and biodiversity affected the poor, who are dependent on naturally available plant species for medicinal purposes (the market value of plants and animals used in the pharmaceutical industry was already $100 billion when the report was published).[92] Worst of all, climate change would impact Sub-Saharan Africa, the Middle East, and South Asia through its impact on water availability, while food yields would be impacted in South America as well.

Today versus tomorrow

Unequal impacts of environmental externalities had consequences on how the case for collective action was calculated. An influential report published in 2006, *The Stern Review on the Economics of Climate Change*, argued that the costs of early action on climate change (1.6 percent of global GDP by 2030) would be outweighed by the significant benefits of avoided losses in future if there were little or no action (upwards of 5 percent of global GDP by 2030).[93] Supporters and critics of the *Stern Review* differed on one crucial point, namely the rate at which the future was discounted. By applying a low discount rate, the *Stern Review* underscored that the lives and life chances of future generations could not be sacrificed to indulge our current lifestyles. This was, indeed, consistent with a human development approach, which would consider shifting the burden of climate change onto future generations ethically indefensible.

At the same time, *HDR 2007/2008* on climate change pointed out that variations in its impacts did not manifest merely over time, but also across geographies. It argued that debates on the discount rate were insufficient as they treated the world as a single country. When the reality of more than 190 countries was considered, the disproportionate impact of climate change on the poorest countries and the poorest citizens within them would not register in global cost-benefit analyses. Distributional issues across generations were different in nature from those within current populations. Whereas the former considered risks of catastrophic losses under uncertainty, the latter would draw attention to the near certain impacts today on some of the world's poorest communities.[94] As climate change impacts were already being felt, a human development perspective would undertake such disaggregated and people-centered calculations, which would not be captured otherwise.[95]

Sustainable consumption

Another fundamental issue was how consumption patterns were undermining the environmental resource base and exacerbating inequalities. The report called for consumption that was shared (to ensure basic needs for all), and that increased human capabilities, and was socially responsible and sustainable.[96] Even with reduced material usage and energy efficiency, pollution and waste were exceeding the Earth's sink capacities while renewable resources (water, soil, biodiversity) were declining in quality as well.[97]

Thus, for developing countries, there was little choice but to adopt a different pattern of development. The report targeted popular myths about the role of developing countries in managing or mismanaging environmental stresses. It argued that rather than expecting developing countries to scale back consumption and industrialization or copy rich countries' environmental policies, there was a case for avoiding the pitfalls of environmental degradation. Leapfrogging to cleaner technologies as well as emission standards could offer a way to continue on a sustainable growth trajectory.[98] But this would mean access to technologies or the financial and technical support to transfer, adapt, and deploy cleaner technologies.

Agenda-setting

In 2007 the Intergovernmental Panel on Climate Change (IPCC) released its Fourth Assessment Report on the science of climate change. It concluded that "it is extremely unlikely that global climate change can be explained without external forcing," implying that emissions of greenhouse gases resulting from human activities had a role to play in driving up average global surface temperatures.[99] Since then the IPCC has published a Fifth Assessment (released in parts in 2013 and 2014). The latter report concluded, "Anthropogenic greenhouse gas emissions have increased since the pre-industrial era … and are *extremely likely* to have been the dominant cause of the observed warming since the mid-20th century."[100]

As the science of climate change became more conclusive, *HDR 2007/2008* addressed the issue by assessing the risks that it posed to continued human progress. It emphasized that despite uncertainty and several "unknowns," the risks of climate change were too high to permit inaction. It argued that uncertainties demanded risk assessments, because the impacts of climate change would threaten future generations with catastrophic outcomes, be irreversible in some regards, and as policy-makers had to account for malign outcomes as much as benign ones.[101] Five risk multipliers were identified: reduced agricultural productivity, greater water insecurity, exposure to coastal flooding and extreme weather events, collapse of ecosystems, and increased health risks.[102] Since then more research has been published on the heightened risks of climate change, permitting scholars based in China, India, the United Kingdom, and the United States to offer a more thorough risk assessment, considering emissions pathways, direct impacts (on food, water, heat, sea levels, and coastlines), and systemic impacts on human social and political systems at national, regional, and global levels.[103]

By establishing that a risk mitigation-oriented approach would draw a common thread across the concerns of rich, developing, and the poorest economies, the human development approach shaped the response to climate change by seeking to influence the agenda on three issues: energy access; technology development and climate financing; and vulnerability and resilience.

Energy access and sustainable energy for all

Energy did not figure prominently in the early discussions on human development. However, *HDR 1998* outlined priorities for sustainable consumption patterns and called for a global goal to provide "access to clean and modern energy services for all," not only for household use but also for communications, transportation, and productive activities. Among other options, it suggested the promotion of decentralized renewable energy, building local capacity to adapt and adopt new energy technologies, and create conditions for increasing rural entrepreneurship in providing energy services at the last mile.[104] These ideas have continued to have resonance as social enterprises have grown, especially in rural areas,[105] and as countries have begun reconsidering their energy strategies.[106]

But energy access did not find any mention in the MDGs. Goal 7 on environmental sustainability had the target of halving the proportion of population without access to safe drinking water or sanitation. It also sought to improve the lives of 100 million slum dwellers and reduce biodiversity loss. However, energy access was glaring in its omission. As a result, discussion on energy access did not emerge in evaluations of how the MDGs were progressing.

HDR 2007/2008 corrected this anomaly and firmly placed energy at the heart of the debate. With 1.6 billion people without electricity (when the report was written) and 2.5 billion dependent on traditional biomass, lack of energy access was retarding human development in a number of areas. Traditional biomass used in cooking was responsible for indoor air pollution killing 1.5 million people every year. It also compounded gender inequalities by increasing the time women had to spend on collecting basic energy resources, rather than spending it on more productive activities. The broader economic costs of low productivity affected households overall, entrenching many in a poverty trap. The demand for energy was also driving deforestation and other environmental effects.[107]

The report bluntly stated, "The vast global deficit in access to basic energy services has to be considered alongside concerns over the rise in CO_2 emissions from developing countries."[108] This also meant that providing access to modern energy services to poor people would not significantly shift the needle upwards on carbon emissions. By corollary, slashing carbon emissions in poor countries would not have as much impact on overall emissions compared with doing the same in rich nations, as the energy footprint of the developed world's population was 20–30 times higher.

In 2004 the United Nations had created a mechanism to coordinate energy-related activities across the UN system. UN-Energy's primary role was to facilitate information sharing and joint programs among UN agencies.[109] In 2007, the UNIDO director-general, Kandeh Yumkella, was elected chair of UN-Energy with a UNDP representative elected co-chair. Access to energy, renewable energy, and energy efficiency became the three pillars of UN-Energy's work.

On 20 December 2010, energy access received a further boost. A UN General Assembly resolution announced 2012 as the International Year of Sustainable Energy for All.[110] With this mandate, the UN secretary-general Ban Ki-moon made sustainable energy for all a key priority for his second term. Specific targets were proposed, including universal access to modern energy services, doubling the rate of improvements in energy efficiency, and doubling the share of renewable energy in the global energy mix by 2030.[111] In April 2011, Kenya and France announced the Paris-Nairobi Climate Initiative, which would focus on energy access activities in Africa and countries most vulnerable to climate change.[112]

Finally, in 2012, the UN Conference on Sustainable Development in Rio de Janeiro (Rio+20) delivered an outcome document, *The Future We Want*, which reaffirmed the critical role of energy in development and poverty reduction.[113] The same document established an intergovernmental process, open to civil society stakeholders, to develop a set of SDGs. The SDGs now include 17 goals and ensuring "access to affordable, reliable, sustainable modern energy services for all" now features prominently. The goal and its underlying targets were adopted at the UN General Assembly in September 2015. New challenges have emerged in the meantime, such as the growing threat of trade disputes over efforts to subsidize clean energy.[114] So, even as energy access is gaining traction within the SDGs and the human development approach, it is not clear yet how other international processes will accept these imperatives and change their rules accordingly.

Technology development, transfer, and financing

The role of technology as a key enabler of human development has been consistently recognized, including in *HDR 2001: Making New Technologies Work for Human Development*. In specific terms, *HDR*s have called for the development and transfer of technologies to confront environmental challenges.

Low-carbon technology is actually shorthand for a swathe of approaches. There is, at a minimum, diffusion of existing technology, with all sectors and countries ratcheting-up to best-practice efficiency levels in order to lower emissions. Then there is accelerating the development and deployment of low-carbon technologies that are at or nearing commercial viability. A third track involves the creation of new breakthrough technologies for achieving zero emissions in power supply and transport, including options ranging from advanced solar power to more embryonic technologies such as nuclear fission.[115]

A year after the Kyoto Protocol was signed (under the UNFCCC), *HDR 1998* called for a shift to renewables, a second green revolution, and removing perverse subsidies. Since then much progress has been made in bringing down the costs of renewable energy or in making fossil fuel-based technologies more efficient. But technology transfer and financing to facilitate such transfer has not always been forthcoming.

Technology transfer has figured on the multilateral agenda for climate change for many years. The UNFCCC called on governments to "take all practicable steps to promote, facilitate and finance, as appropriate, the transfer of, or access to, environmentally sound technologies" (Article 4.5). The Marrakech Accords adopted at the Conference of the Parties (COP-7) in 2001 established a framework for enacting this principle. An Expert Group on Technology Transfer was created to undertake technology needs assessments and identify mechanisms for technology transfer. The Bali "Road Map" called for "enhanced action on technology development and transfer to support action on mitigation." But again and again, climate negotiations offered promises which remained delinked from quantifiable and verifiable outcomes.[116] A more recent effort, the Climate Technology Centre and Network (CTCN), aims at best to provide technical assistance, provide information about climate-friendly technologies, and facilitate collaborations.[117]

Alongside the faltering technology track has been limited climate- and energy-related financing to support developing countries. In less than a decade after the Rio 1992 summit, *HDR 1999* called the Global Environment Facility a "poor cousin" of the ambitious plans set out to generate $125 billion of international resources to help developing countries adopt sustainable development practices. Instead, *HDR 1999* called for a World Environment Agency, which would "oversee the global environment" and report on issues, "broker deals," and "serve as a clearing bank."[118] The clearinghouse functions were particularly important to ensure that developing countries could participate in emission trading schemes but not lose out on long-term access to the global carbon space. It was also necessary to use the new mechanism to generate additional financial resources.

A decade later, in 2009, a similar number ($100 billion) was promised at the Copenhagen Conference of the Parties (COP-15) of the UNFCCC. A mere 10 percent of that sum has been pledged six years hence, with even less disbursed. A new Green Climate Fund was meant to respond to deep governance failures in climate finance, such as greater involvement of developing countries in its governance, increasing the efficiency of disbursement, and improving the credibility with which monitoring occurred.[119] However, without securing additional commitments for finance or raising funds from private capital markets, low-carbon and environmentally friendly technologies would be adopted more slowly than necessary.

Instead, a human development approach to technology development and transfer would consider the drivers of demand for alternative energy technologies. Rather than a top-down imposition of what appropriate technology might be, the approach has to be inversed. There are opportunities to leverage three growing demands: (i) from the poor for access to basic services (and their willingness to pay for the same); (ii) from the middle class for better quality of life and, thereby, efficiency in resource use (adequate energy and water availability, air and water pollution, health impacts, food price inflation); and (iii) from the upper income strata (in developed *and* developing countries) for better returns on investments in technologies and new business opportunities. All the three drivers would expand the choices available to people in rich and poor countries, thereby satisfying the core condition of human development. Thus, the demand for access to basic services could help to develop technology partnerships in renewable energy and decentralized energy. Efficiency of resource use would drive partnerships in energy efficiency (among developed and emerging economies), water use efficiency (particularly among water-stressed and countries with large agricultural sectors), and on air quality standards and monitoring (via partnerships among cities). Finally, the drive for higher returns on investments could promote R&D collaborations on energy storage, cleaner refrigerants, electric and hybrid vehicles, and drought-resistant seeds.[120]

Vulnerability and resilience

More recent *HDR*s have continued to give attention to environmental challenges. But the emphasis has been less on mitigating the problems and more on assessing the scale of impacts and how they might make many communities more vulnerable. The recurring theme across *HDR*s of highlighting the disproportionate impacts on the poor of environmental challenges were again found in *HDR 2011* on sustainability and equity and in *HDR 2014* on reducing vulnerabilities and building resilience.

These were not entirely new ideas. Anand and Sen had argued, "It would be a gross violation of the universalist principle if we were to be obsessed about *inter*generational equity without at the same time seizing the problem of *intra*generational equity."[121] But *HDR 2011* made the relationship between environmental risks and human development more explicit. It found that: (a) household environmental deprivations such as indoor air pollution, or lack of access to water and sanitation were not only severe at low HDI levels but fell as HDI increased; (b) environmental risks with community effects, such as urban air pollution, increased and then fell with rising development; and (c) environmental risks with global effects, that is climate change, rose with HDI.[122] Rather than a direct relationship, it was the overlap of environmental stresses with other human development deprivations (gender-based, education, political participation), which made the relationship complex. Mitigating these challenges meant adopting new strategies to improve resilience to climate change, having robust public-private partnerships, equitable access to finance for climate-related activities (not solely for mitigation), and improved monitoring, verification, and reporting.[123] Once again, these were not new ideas, but they were being placed within a human development framework to ensure both sustainability and equity.

HDR 2014 clearly had the post-2015 development agenda in mind. It was not enough to reduce poverty or improve human development indicators. It was also necessary to ensure that shocks (whether economic, financial, or environmental) did not reverse such progress if development goals had to be genuinely sustainable.[124] On the other hand, an enabling environment was needed to build resilience. Drawing on the multidimensional measurements of human progress (introduced in *HDR 2010*), the report emphasized equal opportunities, lifecycle capabilities, and access to build resilience and empower human agency.[125] In this regard, the *HDR* was not setting the agenda as much as applying the human development paradigm to the agenda, as outlined in *The Future We Want*. From an environmental standpoint, it meant enforcing new templates for disaster prevention and recovery (such as the Hyogo Framework for Action), while continuing to call for strengthening global governance architectures for the provision of global public goods.

Conclusion

In exploring intellectual innovations, policy recommendations, and international negotiations surrounding trade, water, energy, and environment, this chapter finds three common threads running across these themes.

The first is the emphasis on access to opportunity. In international trade, a focus on expanding people's opportunities and choices would mean that the rules would have to be framed in a way that increased access to medicines, or provided rewards for traditional knowledge, or increased capacity for the poorest countries to engage more thoroughly with the global trading system. In water and energy, without access to modern infrastructure, finance, and capacity, many other human development outcomes cannot be achieved.

The second common theme has been on vulnerability and impacts on the poor. International trade is both an enabler of human development goals and could undermine human development progress if the rules are skewed or there are sudden shocks to the trade system. Water and sanitation deficits, or tensions over transboundary waters increase vulnerabilities in lives and livelihoods. Energy access cannot be sacrificed at the altar of responses to global climate change. But, at the same time, environmental externalities gravely impact poor countries and the poorest communities. Unless they are the focus of international and national policy, human development progress will be slow on all counts.

The final common thread is how the intellectual discourses around sustainability and human development have evolved over the past 25 years. As human development has become more explicit about environmental sustainability issues, debates on sustainability have gone beyond mere rhetoric to acknowledge the importance of access, the vulnerability of the poor, increasing resilience, and creating appropriate conditions for developing technologies, and trading and transferring them. In other words, through academic research, policy reports, international negotiations, and declarations of the international community, there is now much more evidence of a growing convergence in policy and practice of sustainable human development.

Notes

1 Richard Jolly, Louis Emmerij, and Thomas G. Weiss, *UN Ideas That Changed the World* (Bloomington: Indiana University Press, 2009), 191.
2 The World Commission on Environment and Development, *Our Common Future* (Oxford: Oxford University Press, 1987).
3 Nitin Desai, "Global Institutions for Sustainability," address to High Level Dialogue on the Institutional Framework for Sustainable Development, Solo City, Indonesia, 19–21 July 2011, 4. Desai served as senior economic advisor to the Brundtland Commission and is credited with developing the definition of sustainable development.
4 Nitin Desai, "Note on the Concept of Sustainable Development," Note to All Commissioners of the World Commission on Environment and Development, 27 June 1986, para. 3.
5 UNDP, *Human Development Report 1994* (New York: Oxford University Press, 1994), 13.
6 Ibid., 18.
7 Dan Bodansky, *The Art and Craft of International Environmental Law* (Cambridge, Mass.: Harvard University Press, 2010), 34.
8 UNDP, *Human Development Report 1994*, 13. Emphasis added.
9 Bodansky, *The Art and Craft of International Environmental Law*, 178, 313, 314. Differential standards and periods of compliance also apply to international trade rules.
10 UN, "United Nations Framework Convention on Climate Change," UN doc. FCCC/INFORMAL/84, 1992, 2. See www.unfccc.int.
11 Ibid., Article 3.1.
12 UN, *Human Development Report 1994*, 19.

13 Ricardo Fuentes-Nieva and Isabel Pereira, "The Disconnect Between Indicators of Sustainability and Human Development," *Human Development Research Paper* 2010/34, 2010, 43.
14 Ibid., 7–10. The authors also discuss Amartya Sen's "sustainable freedom," or preserving capabilities today without compromising the capability to have similar or more freedom in future. Amartya Sen *The Idea of Justice* (Cambridge, Mass.: Harvard University Press, 2009).
15 Sudhir Anand and Amartya Sen, "Human Development and Economic Sustainability," *World Development* 28, no. 12 (2000): 2029–49.
16 UNDP, *Making Global Trade Work for People* (London: Earthscan, 2003), 25–7.
17 Interim Commission for the International Trade Organization, "United Nations Conference on Trade and Employment: Final Act and Related Documents," Havana, Cuba, 21 November 1947–24 March 1948, doc. E/Conf. 2/78, Article 1. www.wto.org/english/docs_e/legal_e/havana_e.pdf.
18 Daniel Drache highlights that labor rights were given equal importance in the Havana Charter along with investment rules. Goals of equity and social justice were paramount. Daniel Drache, "ITO: When Labour and Investment Standards almost Mattered: A Putative History Lesson in Trade Politics that Ought Not to be Forgotten," in *Global Instability: Uncertainty and New Visions in Political Economy*, ed. Stephen McBride, Laurent Dobuzinskis, Marjorie Griffin Cohen, and James Busumtwi-Sam (Amsterdam: Kluwer, 2003).
19 GATT, "General Agreement on Tariffs and Trade: Text of the General Agreement," Article 36, Article 37.1, and Article 18, respectively. www.wto.org/english/docs_e/legal_e/gatt47_e.pdf.
20 WTO, "Agreement Establishing the World Trade Organization," Article 11.2. www.wto.org/english/docs_e/legal_e/04-wto.pdf.
21 Joseph Stiglitz and Andrew Charlton, *Fair Trade for All: How Trade Can Promote Development* (New York: Oxford University Press, 2005), 15.
22 Joseph Stiglitz, *Globalization and Its Discontents* (London: Penguin, 2002), 91.
23 Ha-Joon Chang, *Kicking Away the Ladder: Development Strategy in Historical Perspective* (London: Anthem Press, 2002), 61.
24 UNDP, *Human Development Report 1999* (New York: Oxford University Press, 1999), 2–3.
25 Ibid., 8.
26 Jayshree Watal, *Intellectual Property Rights in the WTO and Developing Countries* (New Delhi: Oxford University Press, 2001), 14.
27 Abdulqawi A. Yusuf, "TRIPS: Background, Principles and General Provisions" in *Intellectual Property and International Trade: The TRIPS Agreement*, ed. Carlos Maria Correa and Abdulqawi A. Yusuf (London: Kluwer Law International, 1998), 8.
28 GATT, Minutes of TRIPS Negotiating Group, doc. MTN.GNG/NG11/11, 14 November 1988, para. 23.
29 GATT, "Meeting of Negotiating Group of 14–15 November 1988," Negotiating Group on Trade-Related Aspects of Intellectual Property Rights, including Trade in Counterfeit Goods, doc. MTN.GNG/NG11/11, 1988, para. 22. https://wto.org/gatt_docs/English/SULPDF/92060156.pdf.

30 Quoted from B.K. Keayla, and Biswajit Dhar, *Issues for Patents (Second Amendment) Bill 1999* (New Delhi: Centre for Study of Global Trade System and Development, 2001), 23 (emphasis added). Prior to 1970 drug prices in India were some of the highest in the world but have since crashed to some of the lowest levels.

31 Oxfam, "South Africa vs the Drug Giants: A Challenge to Affordable Medicines," Oxfam Background Briefing, February 2001.

32 For a brief account of the campaign, see Toby Kasper, "South Africa's Victory for the Developing World," MSF Report, July 2001. www.msfaccess.org/content/south-africas-victory-developing-world, accessed on 30 October 2015.

33 WTO, "Brazil: Measures Affecting Patent Protection—Request for Consultations by the United States," doc. WT/DS199/1, 30 May 2000.

34 In August 2002 the WHO and WTO Secretariat published a joint study, "WTO Agreements and Public Health."

35 WTO, "TRIPS and Public Health—Communication from African Group, Barbados, Bolivia, Brazil, Cuba, Dominican Republic, Ecuador, Honduras, India, Indonesia, Jamaica, Pakistan, Paraguay, Philippines, Peru, Sri Lanka, Thailand and Venezuela," doc. IP/C/W/296, 19 June 2001, para. 5.

36 WTO, "Minutes of TRIPS Council—Special Discussion on Intellectual Property Rights and Access to Medicines," doc. IP/C/M/31, 18–22 June 2001, para. 21–6.

37 The Bush Administration was ideologically committed to launching a new trade round and even offered concessions to publicly skeptical countries like India. See Gary Horlick, "Over the Bump in Doha?" *Journal of International Economic Law* 5, no. 1 (2002): 201.

38 Indian companies were offering generic versions of the antidote ciprofloxacin at 2–5 percent of the cost of the patented drug in the United States.

39 Brazil, Canada, the European Union, India, Kenya, New Zealand, the United States, and Zimbabwe.

40 For a detailed interpretation of the Declaration, see Frederick M. Abbott, "The Doha Declaration on the TRIPS Agreement and Public Health: Lighting a Dark Corner at the WTO," *Journal of International Economic Law* 5, no. 2 (2002): 490–504.

41 Carmen Otero García-Castrillón, "An Approach to the WTO Ministerial Declaration on the TRIPS Agreement and Public Health" *Journal of International Economic Law* 5, no. 1 (2002): 215.

42 WTO, " Decision of 30 August 2003: Implementation of paragraph 6 of the Doha Declaration on the TRIPS Agreement and Public Health," doc. WT/L/540, paragraph 2.

43 Statement by 14 NGOs including Oxfam and MSF, www.oxfam.org.uk/what_we_do/issues/health/wtodeal_300803.htm. Also personal communication from IPA executive, 29 August 2003.

44 UNDP, *Human Development Report 2004* (New York: UNDP, 2004), 88–90.

45 *HDR 2004*, 93.

46 UNDP *Human Development Report 2005* (New York: UNDP, 2005), 126–38.

47 Ibid., 144–5.

48 United Nations, "Resolution adopted by the General Assembly on 16 September 2005," General Assembly, doc. A/RES/60/1, paragraphs 28–31. www.ifrc.org/docs/idrl/I520EN.pdf, accessed on 30 October 2015.
49 WTO, "Aid for Trade," www.wto.org/english/tratop_e/devel_e/a4t_e/aid4trade_e.htm.
50 WTO, "Annual Debate on Aid for Trade—Minutes of Meeting Held in the Centre William Rappard on 21 November 2007," General Council, doc. WT/GC/M/111, 18 December 2007, para. 12, 86–87, 98, 101.
51 WTO, "Follow-Up to the Aid-for-Trade Task Force Recommendations (WT/AFT/1): WTO Monitoring and Evaluation—Report by the Director-General," doc. JOB(06)/262, 12 December, 2006, para. 5–7. WTO, "Agenda Item 8—Aid for Trade—Follow-Up to Report and Recommendations of the Task Force—Statement by the Director-General," doc. JOB(06)/267, 20 December 2006.
52 WTO, "Aid for Trade fact sheet," www.wto.org/english/tratop_e/devel_e/a4t_e/a4t_factsheet_e.htm.
53 Arunabha Ghosh, "Developing Countries in the WTO Trade Policy Review Mechanism," *World Trade Review* 9, no. 3 (2010): 419–55.
54 UNDP, *Human Development Report 1998* (New York: Oxford University Press, 1998), 68.
55 Ibid., 67.
56 UNDP, *Human Development Report 2003* (New York: Oxford University Press, 2003), 31.
57 Ibid., 106–7.
58 UNDP, *Human Development Report 2006* (New York: Palgrave Macmillan, 2006), 1–2.
59 Ibid., 36–7.
60 Ibid., 42.
61 Ibid., 42, 45.
62 Ibid., 49–50.
63 Ibid., 52.
64 United Nations, "International Decade for Action 'Water for Life' 2005–15," www.un.org/waterforlifedecade/background.shtml.
65 United Nations, "Actions taken in organizing the activities of the International Decade for Action, 'Water for Life', 2005–15—Report of the Secretary-General," General Assembly, doc. A/60/158, (25 July 2005). www.un.org/es/comun/docs/?symbol=A/60/158&lang=E.
66 United Nations, "Resolution adopted by the General Assembly on 20 December 2006," General Assembly, doc. A/RES/61/192 (6 February 2007). www.un.org/es/comun/docs/?symbol=A/RES/61/192&lang=E.
67 United Nations, "International Year of Sanitation," http://esa.un.org/iys/ap.shtml.
68 This was in the Dushanbe Declaration on Water, 9 June 2010, www.un.org/es/comun/docs/?symbol=A/65/88&lang=E.
69 *HDR 2006*, 60.
70 United Nations, "Resolution adopted by the General Assembly on 28 July 2010: The human right to water and sanitation," General Assembly, doc. A/RES/64/292 (3 August 2010). www.un.org/es/comun/docs/?symbol=A/RES/64/292&lang=E.

71 UN Human Rights Council, "Resolution 7/22. Human rights and access to safe drinking water and sanitation," (28 March 2008). http://ap.ohchr.org/documents/E/HRC/resolutions/A_HRC_RES_7_22.pdf.
72 WHO, "Drinking-Water, Sanitation and Health," Sixty-Fourth World Health Assembly, doc. WHA64.24 (24 May 2011). http://apps.who.int/gb/ebwha/pdf_files/WHA64/A64_R24-en.pdf.
73 United Nations, "Resolution adopted by the Human Rights Council: The human right to safe drinking water and sanitation," General Assembly, doc. A/HRC/RES/18/1 (12 October 2011). www.un.org/es/comun/docs/?symbol=A/HRC/RES/18/1&lang=E.
74 United Nations, "Report of the Special Rapporteur on the human right to safe drinking water and sanitation, Catarina de Albuquerque," General Assembly, Human Rights Council, doc. A/HRC/24/44 (11 July 2013). www.zaragoza.es/contenidos/medioambiente/onu/1048-eng_report_of_special_rapporteur_on_human_right_to_safe_drinking_water_and_sanitation.pdf.
75 *HDR 2006*, 204.
76 Ibid., 208.
77 Ibid., 210–11.
78 Ibid., 216–17.
79 Ibid., 221.
80 Ibid., 228–31.
81 United Nations, "Water, peace and security: Transboundary water cooperation, Note by the Secretariat," General Assembly, doc. A/64/692 (4 March 2010). Available at: www.un.org/es/comun/docs/?symbol=A/64/692&lang=E.
82 Bodansky, *The Art and Craft of International Environmental Law*, 33 (emphasis in original).
83 Khadija Haq and Richard Jolly, "Global Development, Poverty Alleviation, and North-South Relations," in *Pioneering the Human Development Revolution: An Intellectual Biography of Mahbub ul Haq*, ed. Khadija Haq and Richard Ponzio (New Delhi: Oxford University Press, 2008), 82.
84 UNDP, *Human Development Report 1990* (New York: Oxford University Press, 1990), 13, 22, 36, 38.
85 UNDP, *Human Development Report 1992* (New York: Oxford University Press, 1992), 18.
86 Ibid., 83–4.
87 *HDR 1994*, 23.
88 Ibid., 28–9, 35–6.
89 *HDR 1998*, 66.
90 Ibid., 68–9.
91 Arunabha Ghosh and Hem Dholakia, "Competitive Pollutionism No Good" *Business Standard*, 17 March 2015. www.business-standard.com/article/opinion/arunabha-ghosh-hem-himanshu-dholakia-competitive-pollutionism-no-good-115031601230_1.html.
92 *HDR 1998*, 75.
93 Nicholas Stern, *The Economics of Climate Change: The Stern Review* (Cambridge: Cambridge University Press, 2006).
94 Partha Dasgupta, *Human Well-Being and the Natural Environment* (Oxford: Oxford University Press, 2001).

95 UNDP, *Human Development Report 2007/8* (New York: Palgrave Macmillan, 2007/8), 64.
96 *HDR 1998*, 1.
97 Ibid., 4.
98 Ibid., 84.
99 IPCC, "Summary for Policymakers," in *Climate Change 2007: The Physical Science Basis. Contribution of Working Group I to the Fourth Assessment Report of the Intergovernmental Panel on Climate Change*, ed. S. Solomon, D. Qin, M. Manning, Z. Chen, M. Marquis, K.B. Averyt, M. Tignor and H.L. Miller (Cambridge: Cambridge University Press, 2007).
100 IPCC, "Summary for Policymakers," in *Climate Change 2014: Synthesis Report. Contribution of Working Groups I, II and III to the Fifth Assessment Report of the Intergovernmental Panel on Climate Change*, ed. R.K. Pachauri and L.A. Meyer (Geneva: IPCC, 2014), 4.
101 *HDR 2007/2008*, 58–9.
102 Ibid., 27, 30.
103 David King, Daniel Schrag, Zhou Dadi, Qi Ye, and Arunabha Ghosh, *Climate Change: A Risk Assessment* (London: UK Foreign and Commonwealth Office, July 2015).
104 *HDR 1998*, 88.
105 In India alone there are now more than 250 firms operating in the decentralized clean energy sector. Arunabha Ghosh, Rajeev Palakshappa, Rishabh Jain, and Poulami Choudhury, "Developing Effective Networks for Energy Access: An Analysis," Washington, DC: US Agency for International Development, August 2013, 6.
106 Arunabha Ghosh and Karthik Ganesan, "Rethink India's Energy Strategy," *Nature* 521 (14 May 2015): 156–7. www.nature.com/news/policy-rethink-india-s-energy-strategy-1.17508.
107 *HDR 2007/2008*, 45.
108 Ibid., 44.
109 UN-Energy, "About UN-Energy," www.un-energy.org/about/overview.
110 United Nations, "Resolution adopted by the General Assembly: International Year of Sustainable Energy for All," General Assembly, doc. A/RES/65/151 (16 February 2011). www.un.org/ga/search/view_doc.asp?symbol=A/RES/65/151.
111 Sustainable Energy for All, www.se4all.org/.
112 Paris-Nairobi Climate Initiative, www.developpement-durable.gouv.fr/IMG/pdf/The_Paris-Nairobi_Climate_Initiative.pdf
113 United Nations, "Resolution adopted by the General Assembly on 27 July 2012: The future we want," General Assembly, doc. A/RES/66/288 (11 September 2012). www.un.org/ga/search/view_doc.asp?symbol=A/RES/66/288&Lang=E.
114 Arunabha Ghosh with Himani Gangania, *Governing Clean Energy Subsidies: What, Why, and How Legal?* (Geneva: International Centre for Trade and Sustainable Development, 2012).
115 Nicholas Stern, *A Blueprint for a Safer Planet: How to Manage Climate Change and Create a New Era of Progress and Prosperity* (London: The Bodley Head, 2009), 111–15, 147; Nicholas Stern, "Key Elements of a Global Deal on Climate Change," London School of Economics and

Political Science, 2008, 32–4; Arunabha Ghosh and Kevin Watkins, "Avoiding Dangerous Climate Change: Why Financing for Technology Transfer Matters," Global Economic Governance Working Paper 2009/53, Oxford, 2009, 21.

116 Ghosh and Watkins, "Avoiding Dangerous Climate Change," 37. Also, see Arunabha Ghosh, Anupama Vijayakumar, and Sudatta Ray "Climate Technology Partnerships: Form, Function and Impact," *CIGI Fixing Climate Governance Series – Paper No. 2*, October (Waterloo, ON: Centre for International Governance Innovation, 2015). www.cigionline.org/publications/climate-technology-partnerships-form-function-and-impact.

117 Climate Technology Centre & Network, "About the Climate Technology Centre & Network (CTCN)," http://ctc-n.org/about-ctcn.

118 *HDR 1999*, 113.

119 Arunabha Ghosh, "Harnessing the Power Shift: Governance Options for International Climate Financing," Oxfam Research Report, Oxfam International, Oxford, October 2010, 1–90.

120 Arunabha Ghosh, "Making the UN Secretary General's Climate Summit Count: An Opportunity for Credible Climate Leadership," CEEW Policy Brief, September 2014, 4–7. http://ceew.in/pdf/ceew-issue-brief-making-the-un-secretary-general-climate-summit-count-17sep14.pdf.

121 Sudhir Anand and Amartya Sen, "Human Development and Economic Sustainability," *World Development* 28, no. 12 (2000): 2038.

122 UNDP, *Human Development Report 2011* (New York: Palgrave Macmillan, 2011), 3.

123 Ibid., 12.

124 UNDP, *Human Development Report 2014* (New York: UNDP, 2014), 2.

125 Ibid., 4.

5 Human development in international policy-making, Part II

Democratic governance, human rights, and peacebuilding

- **Democratic governance and human rights**
- **Peacebuilding**
- **Conclusion**

Similar to human development's critical relationship with trade, water, energy, and the environment, introduced in Chapter 4, democratic governance, human rights, and peacebuilding are each fundamental to both the evolution and continued policy relevance of human development. Together, their diagnosis in connection with multiple global, regional, national, and sub-national *Human Development Reports* (*HDR*s) have steered the human development discourse from beyond its traditional moorings and comfort zone in the disciplines of development economics, sociology, and women's studies. This chapter considers human development's contribution to: a paradigmatic intellectual shift; an influential international policy agenda-setting role; and innovative reforms to the changing nature of democracy, human rights, and peacebuilding in global and national governance. In doing so, it seeks to demonstrate how human development has benefited from the analytical tools of political scientists, lawyers, and philosophers, thereby enriching the concept of human development and its policy relevance in global institutions.

Democratic governance and human rights

Despite successive waves of democratic development in recent decades—beginning with southern Europe in the 1970s, Latin America in the 1980s, and Eastern Europe and large parts of Asia in the 1990s—the period since 2000 has witnessed a retreat in influential policy circles about the effectiveness of democratic governance to deliver and sustain economic and social progress.[1] Still, multilateral institutions, from the

United Nations to the Organization for Security and Co-operation in Europe (OSCE) and the Organization of American States (OAS), continue to assume growing responsibilities and resources for the promotion of democracy, particularly in fragile and conflict-affected states.

Meanwhile, the UN Human Rights Commission was transformed into a new Human Rights Council in 2005, with a new set of tools and procedures to safeguard human rights internationally. However, the forces of globalization and inherent weakness in the present system of global governance introduced in Chapter 1 raise doubts about the future of democratic governance and human rights—two concepts central to the rise of human development since 1990 as a chief international policy concern.

Whereas human development involves the enlarging of people's choices to improve the human condition, democracy involves expanding people's choices about how and by whom they are governed. In doing so, democracy brings participation, accountability, and other principles to the process of human development. Similarly, human rights intersect with and enrich human development by safeguarding people's political, civil, economic, social, and cultural rights. Without democratic forms of governance and the promotion of human rights, human development is more than impaired; it lacks an essential core. And without a clear embrace and progress in advancing broad human development policy priorities in both global and national governance, democratic governance and human rights cannot be sustained.

Paradigm shift?

Democracy is associated with the idea of "the will of the people," which derives from the centuries long transition from "absolute sovereignty"—examined in Thomas Hobbes' classic study *Leviathan*—to "popular sovereignty," where the people rule themselves. State authority regulated by institutional checks and balances (constitutional authority) and democratic norms, such as popular participation and the peaceful contestation of power, were innovated in the Greek city-state of Athens and later by political philosophers, such as John Locke and Jean-Jacques Rousseau. "To understand Political Power right, and derive it from its Original, we must consider what State all Men are naturally in, and that is, a State of perfect Freedom to order their Actions … without asking leave, or depending upon the Will of any other Man," writes John Locke.[2] Christopher McMahon and other modern-day political philosophers argue that democracy possesses greater moral validity or legitimacy than any other way of exercising authority,[3] and according to William Connolly, "The genius of democracy is that it allows social conflicts to find open expression, moderates the intensity of those conflicts, and provides procedures by which to legitimize public resolution."[4] Consequently, democracy at least maintains the potential to create favorable conditions for the concept and policies linked to human development to take root and flourish.

Similarly, the modern definition and application of human rights is both consistent with and reinforcing of current conceptions and efforts to "operationalize" human development. In sum, human rights express the bold idea that all people have claims to social arrangements that protect them from the worst abuses and deprivations, as well as secure human freedoms in seven fundamental areas:

- Freedom from discrimination—for equality.
- Freedom from want—for a decent standard of living.
- Freedom for the realization of one's human potential.
- Freedom from fear—with no threats to personal security.
- Freedom from injustice.
- Freedom of participation, expression, and association.
- Freedom for decent work—without exploitation.[5]

From the Universal Declaration of Human Rights (1948), to the International Covenant on Civil and Political Rights and International Covenant on Economic, Social, and Cultural Rights (both 1976), tremendous strides were made in the twentieth century in the promotion and protection of human rights worldwide. With these advances, human development—which is also committed to securing basic freedoms—also benefited through the enhancement of human capabilities of ever-growing national populations, thereby further enlarging their choices at the very heart of human development.

From a conceptual standpoint, the scholars and policy analysts preoccupied with democracy and human rights, on one hand, and human development, on the other, hail from distinct traditions and divergent strategies of analysis and action. But their parallel paths began to converge in the 1990s in the aftermath of the Cold War. The political scientists, lawyers, and philosophers advocating for the concepts of democracy and human rights have placed the missing dimension of politics squarely at the center of human development, with the growing recognition that reducing poverty depends as much on whether poor people have political power as on their opportunities for economic progress. These scholars and analysts add value to the broader development agenda by drawing attention to accountability and the fulfillment of human rights for all people. They further bring legal tools and institutions—laws, the judiciary, and the process of litigation—as effective and sometimes innovative means to secure basic freedoms and human development.[6] Moreover, they incorporate the principle of social justice and lend moral legitimacy to human development objectives.

At the same time, human development enriches thinking on democracy and human rights. First, it focuses attention on the often less understood social and economic dimensions of democratic theory and practice, which if not effectively addressed can undermine democratic political participation in a national polity.[7] Similarly, as noted in the next section, human development has established greater parity since the early 1990s between economic, social, and cultural rights, on one hand, and political and civil rights, on the other. Through the introduction of new concepts, tools, and resources, it provides a systematic assessment of economic and institutional constraints to the realization of all kinds of rights.[8] And in an era when people and governments demand significant progress in real time, human development has helped to temper expectations by bringing a much-needed long-term perspective to the promise of both democracy and human rights.

Within global governance too, human development has intersected with both democracy and human rights in innovative ways to push the boundaries of both sets of concepts. With its emphasis on people as agents for progressive change, the *HDR*s have reinforced the case for "increasing pluralism," by steadily expanding the space and influence of nonstate actors in global decision-making, including through the development of new mechanisms to moderate the behavior of private corporations.[9] This is no small feat given the continued sacrosanct view with which many countries continue to hold on to the notion of absolute sovereignty as the bedrock of the international system.

Simultaneously, the *HDR*s have sought to give countries from the Global South a larger role in global decision-making by increasing participation and accountability in multilateral institutions, as well as by leveraging the established body of human rights laws, norms, and instruments in a manner that steers greater political attention and resources to the human development priorities of developing countries. For instance, the special rapporteurs and independent experts established by the United Nations to focus on development concerns, including neglected economic, social, and cultural rights, have brought greater attention to major challenges where a great majority of the world's population resides. In short, the end of the Cold War created the conditions for both a conceptual and, as shown below, policy convergence between the previously separate development and democracy-human-rights communities. This intersection has contributed to conceptual innovations in the human development school of thought and, in many ways, a paradigmatic shift in how one thinks about development, democracy, and human rights.

Agenda-setting?

Beginning in the 1990s, the coming together of thinkers and activists on development, democracy, and human rights through the vehicle of global, regional, national, and sub-national *HDR*s contributed to advances in international policy-making reforms and institution-building in three ways. First, by encouraging the recent waves of democratization, while stressing that democracies can only survive—let alone thrive—when governments and societies invest in people. Second, by lending support to civil-society-championed global governance innovations, from the International Criminal Court to the Jubilee 2000 Debt Relief and Ban Landmine campaigns. Third, by strengthening the tools and increasing resources for promoting democracy and human rights as integral elements of human development.

Investing in people as central to fostering and sustaining democratic development

Stressing the centrality of human agency, the inaugural global *Human Development Report 1990* underscored the importance for governments and donors alike to invest in the skills and aspirations of people to both further and safeguard recent achievements globally in the expansion of human freedoms. Alongside advocating for the need for citizens to "have constant access to decision-making and power," *HDR 1993*, for instance, offered a checklist for effective participation that included: equitable access to health and other aspects of physical well-being; equitable access to knowledge, skills, technology, and information; and equal human rights.[10] In subsequent global *HDR*s—dedicated to the themes of empowering women (1995), eradicating poverty (1996), human rights (2000), and deepening democracy (2002)—the case for targeted investments in people, especially the most vulnerable in society, was carefully laid out. Strengthening human agency, these studies argued, was not only beneficial to the individuals involved but fundamental to building strong and free societies.

At the same time, individual agency was, in part, contingent on recognition and accommodation of their identities, whether based on gender, religion, language, ethnicity or color, and so forth. Human development rejects cultural determinism, because to do so would be to deny individuals the power to shape their own destinies based on their own expanding choices and capabilities. Yet, if cultural attributes became a reason for exclusion of individuals or whole groups of people (exclusions based on lifestyles or on broader political, social, and economic participation), then progress in human development could be curtailed or even reversed. Thus, *HDR 2004* pushed the boundaries of democracy, human rights, and human development by arguing in favor of protecting and strengthening cultural liberties. The report was not a celebration of the diversity in human cultures, but the liberties that must be entailed within multi-ethnic, multireligious, and multilingual societies for individuals to draw on their multiple identities. Put simply, "cultural liberty is central to the capability of people to live as they would like."[11]

Balancing individual rights, choices, and freedoms with recognition of community attributes was not going to be an easy task for democracies. Measures relating to ensuring wider political participation (how would minorities be treated within the constitution of a newly independent state, for instance), ensuring religious freedom, permitting legal pluralism, language policies, and policies targeting structural inequities all were discussed in the report. A human development perspective on cultural liberties ensured that individual rights would still need to be protected by state policy, rather than defend traditional practices, which violated human rights. The same logic was extended to knotty issues of how to engage with extremist ideologies and violent movements, the role of indigenous knowledge (see Chapter 4), other cultural products, and finally, migration and integration into host societies.

Although it was nearly impossible to introduce language on democratization in UN resolutions and documents during the Cold War, the waning years and period that followed were ripe for multiple concurrent international dialogues both parallel to and within the UN. In 1988, the first in a series of International Conferences on New and Restored Democracies closely affiliated with the UN was convened, and in 2000, the United States encouraged the formation of a "Community of Democracies" to strengthen democracy worldwide by providing support to emerging democracies and civil society. In the same year, a breakthrough was achieved at the UN, when the Millennium Declaration, signed by 147 heads of state and government, proclaimed that "We will spare no effort to promote democracy and strengthen the rule of law, as well as respect for all internationally recognized human rights and fundamental freedoms, including the right to development."[12] And in 2007, the UN General Assembly declared 15 September as the International Day of Democracy, while simultaneously stressing that "democracy, development and respect for all human rights and fundamental freedoms are interdependent and mutually reinforcing."[13] Throughout the 1990s and 2000s, the *HDR*s provided a high-profile, steady drumbeat in support of these democratization support efforts, while ensuring that these forums and documents also afforded attention to the need for democratic governments to demonstrate socio-economic development outcomes through strategic investments in their most important national asset: their citizens.

The *HDR*s have consistently reinforced the argument that for democracies to survive, let alone thrive, their governments must demonstrate performance and respond to the growing demands and aspirations of their citizens. Failure to do so places experiments in democracy at risk, as shown over the past decade by the rise of right-wing populism in large parts of Europe and the renewal of left-wing populism across Latin America. *The Economist* attributes growing disenchantment with democracy and a measured decline in global freedoms by Freedom House and other sources to the financial crisis of 2007–08 and the rise of China, where "the Chinese Communist Party has broken the democratic world's monopoly on economic progress."[14]

In many ways, building the institutions needed to sustain democracy is more difficult than drafting a new constitution and convening elections. Ample evidence also suggests that nurturing democratic practice requires a strong or at least aspiring middle class. By equipping policy-makers with the analytical tools and role model stories on successful democratic governance that simultaneously privilege social-economic development, human development continues to make a significant contribution to the institutions and practice of democratic development.

Reinforcing global governance reforms endorsed by global civil society

Another policy area where the *HDR*s have placed a spotlight on the democratization of global governance is in showcasing the influential role of civil society organizations acting in concert globally. From the earliest *HDR*s, the substantial contributions of civil society, including international nongovernmental organizations, to addressing a range of national and local human development challenges were documented and examined through a critical lens. But the *HDR 2002* represented a breakthrough in underscoring the vital role of civil society campaigns and associated "multi-stakeholder processes," involving governments, international organizations, and business groups, in injecting greater pluralism into international decision-making, thereby strengthening global democracy.

Three issues stand out. First is the *Human Development Report 2002*'s analysis of the impact of the Jubilee 2000 campaign in winning debt relief concessions, including an expanded heavily indebted poor countries initiative and bilateral commitments by donor countries to write off debt from developing countries. Second is the campaign for an International Criminal Court (ICC), which was established following the sixtieth ratification in April 2002 of the court's founding treaty, the Rome Statute. Third is the International Campaign to Ban Landmines, which in concert with a coalition of like-minded countries brought about the signing in 1997 of the Ottawa Treaty—the Anti-Personnel Mine Ban Convention, or Mine Ban Treaty. The latter was later ratified and became binding international law in 1999. Coming off the profound impact of anti-globalization protesters at the 1999 World Trade Organization Ministerial Conference in Seattle, the *HDR*s joined a growing chorus of voices in demonstrating the tangible contributions of civil society actors to international policy development.

Among the implementation challenges faced subsequently by these global governance innovations is the criticism that the International Criminal Court has only addressed cases from Africa and that 35 countries still refuse to sign the Mine Ban Treaty, including the United States, Russia, China, Israel, and Iran. Meanwhile, another major civil-society-championed global norm, the Responsibility to Protect—enshrined in the 2005 UN Summit Outcome document—has been rebuffed in connection with the on-going conflict in Syria, despite its application in justifying airstrikes in 2011 in Libya by the North Atlantic Treaty Organization (NATO). Nevertheless, the *HDRs* have taken a stand in support of a multitude of influential global governance reforms endorsed by civil society over the past two and a half decades, including in the promotion of democratic principles in the United Nations, World Bank, International Monetary Fund, and World Trade Organization.

Building on the 1995 Commission on Global Governance, the 1995 Independent Working Group on the Future of the United Nations, and earlier global commissions and independent studies on the future of the international system, successive *HDR*s in the 1990s and 2000s advocated for the following reforms:

- Increasing civil society organization participation in the discussions of the General Assembly, Economic and Social Council, and

Security Council, as well as supporting calls for direct citizen participation in the work of the UN system;

- Expanding the global economic decision-making clout of developing countries by transforming ECOSOC into a powerful 22-member Economic Security Council, including 11 permanent members from the main industrial and more populous developing countries.
- Making the World Bank and IMF more accountable for their actions to board members but also to the people affected by their actions.
- Improving World Trade Organization consultations, discussions, negotiations, and decision-making so as to make the WTO more transparent, participatory, and democratic.
- Expanding the composition of the UN Security Council and modifying the use of the veto by its Permanent Five (P-5) members (China, France, Russia, the United Kingdom, and the United States).

Besides providing creative reform proposals for civil society groups and like-minded governments seeking to strengthen global governance, several policy prescriptions have garnered some traction. For example, civil society organizations have steadily expanded their political reach and formal channels of participation in major UN organs, and the Group of 20—meeting at the heads of state level since the 2008–2009 global financial crisis—has key global economic governance features associated with the Economic Security Council proposal. At the same time, proposed changes to the composition of the Security Council and use of veto by the P-5 continue to meet resistance.

Strengthening the instruments and resources for democracy and human rights promotion

A third and final area where the reports have shaped international policy-making reforms and institution-building involves the careful application of the tools and resources required for the effective promotion of democracy and human rights. Much more than a technical exercise, the *HDR*s highlight the political skills and contested approaches associated with the burgeoning fields of democratic governance and human rights assistance. The *Afghanistan Human Development Report 2007*, for example, challenges the rapid introduction of Western models of justice in Afghanistan, while recommending a "Hybrid Model of Formal and Informal Justice" for a transitional period which could expand access to justice through a partial acceptance of traditional local *jirgas* for dispute settlement.[15] Following an endorsement of this approach by the Afghan Ministry of Justice in 2008, the United States Agency for International Development (USAID) funded a pilot program in several districts across Afghanistan. Similar questions had been discussed in *HDR 2004*, such as power sharing arrangements in post-conflict societies, federalism, self-rule for First Nation communities, or plural legal systems and customary law. Beyond investing in modern institutions of democratic governance, the *Kosovo Human Development Report 2004* stressed the importance of family and community-level networks, as well as the rise of nongovernmental organizations, in helping to promote human security, deepen democracy, and protect the rights of all Kosovans.[16]

Beyond specific recommendations considered by national and international policy-makers, the *HDR*s introduced new analytical and measurement tools to assess progress in the areas of democratic development and human rights promotion. A Human Freedom Index was introduced in *HDR 1991*, followed by the introduction of a, similar, Political Freedom Index for *HDR 1992*.[17] However, these composite indicators were widely criticized by UN member states which ranked poorly, and they were subsequently discontinued.[18] It was not until the global reports on human rights (2000) and democracy (2002) that data on political and civic freedoms were widely collected and introduced as indicators for assessing progress and ensuring accountability. Alongside the pioneering Human Development Index, Human Poverty Indices, and Gender-Related Development Index (see Chapter 3 of this volume) innovated in the 1990s, *HDR 2000* and *HDR 2002* demonstrated that democratic development and human rights could be measured in acceptable ways internationally. This has enriched the concept and practice of human development in all societies, whether rich or poor or from the West or East.

At the same time, as highlighted in the 1994, 2002, and 2005 *HDR*s, and further elaborated in the section below on peacebuilding and human development, the path for countries transitioning from authoritarian regimes and violent conflict to more democratic and peaceful forms of governance is rarely linear and can be fraught with challenges. In a major 1995 study, for example, Edward Mansfield and Jack Snyder found that while mature, stable democracies are safe, states " … usually go through a rocky transition [to democracy] … Statistical evidence covering the past two centuries shows that in this transitional phase of democratization, countries become more aggressive and war-prone, not less, and they do fight wars with democratic states."[19] This is, in part, why professionalizing a country's security forces and placing them quickly under the oversight of elected, civilian leaders is critical to a nascent democracy's development.

Similarly, the leaders of many state and nonstate actors worldwide continue with impunity—as manifested in daily news reports—to express a blatant disregard for accepted international human rights norms. This despite refined tools such as human rights indicators and monitoring, as well as new institutions from the ICC to the UN Human Rights Council, for holding human rights violators to account. In short, a solid understanding of a national context and a good sense of timing also are essential for maximizing the impact of both tools and resources allocated to democracy and human rights promotion.

New horizon issues

With the above conceptual innovations and policy reform ideas serving as a starting point, two "new horizon" issues are identified below. Detailed further in Chapter 6, they seek to strengthen the intersection between democracy, human rights, and human development in global governance.

The first is enhancing global governance tools for democratic governance promotion at the national and sub-national levels. Encouraging and sustaining democratic transitions are a fundamental challenge worldwide, and the UN has only just begun to invest in the platforms and instruments for strengthening and sustaining national and sub-national democratic governance. The UN requires innovative tools to build and sustain global networks for democratic change. This includes revitalizing the International Conferences on New and Restored Democracies (which the UN began to support directly in 1994); and increasing the resources, technical assistance, and studies provided through the UN Democracy Fund, UN Department of Political Affairs, UN Development Programme, and UN University.

Although the effort may be global in nature, the UN's focus should be on deepening democratic institutions and practices within nation-states, especially in connection with sub-national geographic units and minority groups that may feel excluded from a country's political, economic, and social progress. This is also the essence of a human development approach to engendering more inclusive and accountable democratic systems of governance.

Second, is the need to place economic, social, and cultural rights on a par with political and civil rights internationally. Despite coming into force simultaneously with the International Covenant on Civil and

Political Rights in 1976, the International Covenant on Economic, Social, and Cultural Rights has not received the same high-level political attention since its inception. Although innovations have helped to place economic, social, and cultural rights on the international agenda, these efforts remain inadequate. These include initiatives such as the UN "independent experts" on human rights and extreme poverty (1998) and on the effects of economic reform policies and foreign debt on the full enjoyment of human rights (2000); as well as a "special rapporteur on economic, social, and cultural rights," and special rapporteurs on specific issues such as the human right to education (1998), adequate housing (2000), food (2000), and safe drinking water and sanitation (2008). A human development approach places economic, social, and cultural rights on an even par with political and civil rights in global governance, which would imply, for instance, that the UN High Commissioner for Human Rights and the UN Human Rights Council offer similar resources and high-level political attention to these central aspects of the global rights agenda.

Peacebuilding

With an estimated 1.5 billion people living in fragile and conflict-affected countries and the fact that no recent conflict-affected state has achieved a single Millennium Development Goal,[20] peacebuilding matters to human development. Most of the citizens in these countries lack basic human development opportunities, in part because of uncertainties around the state's ability to ensure the safety and protection of its people. When individuals are consumed with fundamental issues of survival and personal security, planning for and investing in the future takes a back seat. In short, peacebuilding provides local actors—both state and nonstate—with the tools for managing and addressing the sources of violent conflict, thereby creating the conditions for human development to flourish.

At the same time, human development has enriched thinking and action on peacebuilding in two major ways. First, by demonstrating how building durable peace requires people's empowerment and the development of inclusive institutions capable of safeguarding their rights, as detailed earlier in this chapter. Second, through the corollary concept of human security, the *HDR*s series analyzes the risks and threats to human development when the security of people is not balanced in relation to the security of states. Besides contributing to a paradigmatic shift in how one understands the causes of violent conflict, human development and human security have expanded the peacebuilding toolbox for mitigating and neutralizing the drivers of deadly political and criminal violence, including through an emphasis on preventive development and holding state security institutions accountable to elected civilian leaders. Despite some notable successes since the end of the Cold War, pursuing this agenda has faced and continues to face innumerable challenges and constraints.

Paradigm shift?

The Norwegian scholar Johan Galtung first coined the term peacebuilding in 1975, defining peacebuilding as a means of "preventing incipient violent conflicts by addressing root causes of poverty, political repression and uneven distribution of resources."[21] In his 1992 *An Agenda for Peace*, then-UN secretary-general Boutros Boutros-Ghali both narrowed and deepened the concept when he wrote that "post-conflict peacebuilding" refers to "action to identify and support structures which will tend to strengthen and solidify peace in order to avoid a relapse into conflict."[22] Alongside the related concepts of preventive diplomacy, peacemaking, and peacekeeping, post-conflict peacebuilding helped to reinvigorate support for international, and particularly UN, responses to protracted armed conflict both between and within states. From 1992 to 1996, UN peace operation and conflict management budgets and staffing would witness the first of two growth spurts in the 1990s and later in the 2000s, leading to engagements heretofore unprecedented in size, scope, and level of political intrusiveness.

As noted above and earlier in Chapter 2, human security, introduced in *HDR 1993* and subsequently elaborated in the *HDR 1994*, also shaped how international policy-makers view the origins of and an appropriate response to deadly armed violence. Defining human security as "people exercising their human development choices safely and freely" and "ensuring freedom from fear and freedom from want,"[23] the *HDR*s immediately contributed to a broader understanding of the sources of violent conflict and the need to inject targeted resources for human development into any kind of strategic framework or individual (sectoral) measures aiming to cope with violent conflict and reduce the likelihood of its recurrence. In particular, human development and human security placed people at the center of future conceptions of peacebuilding, serving as both the ends and chief means towards achieving sustainable peace. Consequently, they markedly changed the international mindset on the requirements for effective peacebuilding, freed in part by the lapse in the superpower (and modern military and state-centric) standoff that virtually paralyzed global institutions such as the UN during the more than four decades long Cold War

Beginning in the 1990s and steadily growing in significance during the first decade of the twenty-first century, development and, more specifically, human security-related concepts and activities made their way into UN and wider international policy statements, and informed actions on conflict mitigation and response. For example, in 1999 the government of Japan and the United Nations Secretariat launched the United Nations Trust Fund for Human Security, which finances activities carried out by UN agencies and designated non-UN organizations that seek to translate the human security approach into practical, field-level actions. At the 2005 UN Summit, in the outcome document's "Human Security" section, the heads of state and government stressed "the right of all people to live in freedom and dignity, free from poverty and despair," and recognized that "all individuals, in particular vulnerable people, are entitled to freedom from fear and freedom from want, with an equal opportunity to enjoy all their rights and fully develop their human potential."[24]

Also in the outcome document, world leaders endorsed for the first time the norm of the "Responsibility to Protect" (R2P) populations from genocide, war crimes, crimes against humanity, and ethnic cleansing.[25] While the UN-sanctioned NATO air and sea mission in Libya (2011), and French and UN peacekeeping intervention in the Central African Republic (2013–14), represent applications of R2P, the failure of the UN Security Council (resulting, in part, from vetoes by Russia and China on successive resolutions) to prevent more than 200,000 civilian and combatant deaths in Syria (2011–present) demonstrates the challenge to consistent, universal acceptance of this new international norm.

In sum, human development and human security informed the evolution of the similarly complex and integrative concept of peacebuilding, paving the way for several innovative policy responses (see below). At the same time, the experience and knowledge generated through the policy and practice of peacebuilding over the past two decades has further enhanced the meaning of human development and how its various applications are understood in global governance.

Agenda-setting?

As seen through the lens of human development and manifested in the writings of the global *HDR*s and related human development publications, peacebuilding has contributed to international agenda-setting in at least three concrete areas. It has shifted the focus of international policy-makers from a predominant focus on the security of states to a balanced emphasis on both the security of states and their citizens. It has introduced both conflict analysis and sensitivity into the international development discourse. And finally it has informed the need and design of the new "Global Peacebuilding Architecture," including the UN Peacebuilding Commission (PBC), Peacebuilding Support Office, and Peacebuilding Fund.

From security of states to security of both states and people

In 2007, the UN Secretary-General's Policy Committee agreed on the following conceptual basis for peacebuilding to inform UN practice:

> Peacebuilding involves a range of measures targeted to reduce the risk of lapsing or relapsing into conflict by strengthening national capacities at all levels for conflict management, and to lay the foundations for sustainable peace and development. Peacebuilding strategies must be coherent and tailored to specific needs of the country concerned, based on national ownership, and should comprise a carefully prioritized, sequenced, and therefore relatively narrow set of activities aimed at achieving the above objectives.

This broad conception of peacebuilding—subsequently enshrined in 2009 in the *Report of the Secretary-General on Peacebuilding in the Immediate Aftermath of Conflict* and now considered UN orthodoxy—stems directly from the earliest thinking on human security from *HDR 1994* with its emphasis on people as the chief agents for achieving sustainable peace and development. In this sense, subsequent international policy discussions on the conflict-security-development nexus, including calls for a Global Peace Dividend Fund, in the lead up to the 1995 World Summit for Social Development set the stage for peacebuilding-related policy debates in the 2000s at the international level.

As noted above, by 1999, the UN had established a new Trust Fund for Human Security to translate the human security approach into practical, field-level actions, and many lessons from this trust fund were subsequently channeled to a new Peacebuilding Fund set up in 2005 as an integral part of the new Global Peacebuilding Architecture (see below). Informing today's approach to peacebuilding, the UN Trust Fund for Human Security has contributed for over a decade to "rebuilding war-torn communities; protecting people exposed to extreme poverty, sudden economic down-turns, and natural disasters; and addressing urban violence, trafficking-in-persons, arms, and illicit substances; among others."[26] In addition, human security and peacebuilding would prove instrumental in stressing the importance of development—and the central role of people as drivers of their own development—to conflict prevention and management in several influential UN reports in the early 2000s, including the 2000 *Report on United Nations Peace Operations* (Brahimi Report); the 2001 *Report of the Secretary-General on the Prevention of Armed Conflict*; the 2004 *Assessment of the Ad Hoc Advisory Group of the Economic and Social Council on African Countries Emerging from Conflict*; and the 2004 *Report of the Secretary-General's High-Level Panel on Threats, Challenges, and Change*. At the same time, it should be acknowledged that although foreign aid levels among traditional OECD donors, including the United States, and new emerging donors, including China and South Korea, grew significantly over the past decade, their military expenditure levels rose also, manifesting a continued reliance on twentieth-century (traditional) means for promoting state security.

Conflict analysis and sensitivity

Beginning with the introduction of the concept of human security in the early 1990s, the *HDR*s sensitized policy-makers to the need to bring conflict analysis into decision-making about development policies and programs. At the very least, development decisions would aspire to employ a "do no harm" principle in conflict-affected countries, but ideally, development activities would serve to build national capacities for improved conflict management. This point is manifested in a reference to the concept of peacebuilding in *HDR 2002*:

> Securing a just, sustainable peace in conflict-prone situations means building strong, transparent states with professional, civilian-led military and police. It means developing a democratic framework that tolerates diversity. It means building an open civil society that promotes democratic governance and personal security. And it means instilling in all state institutions—but especially the security forces—a culture of democracy rooted in respect for the rule of law and individual rights and dignity. This is the essence of democratic peacebuilding.

The treatment of security sector reform and, in particular, democratic civilian oversight of a country's armed forces would soon become a mainstream activity in the international development community, including through guidelines such as those on Security System Reform and Governance in 2005 by the OECD Development Assistance Committee.

In a chapter on "violent conflict—bringing the real threat into focus," *HDR 2005* both captured and advocated for new approaches to aid for "conflict-sensitive development." In concrete terms, it called for governments, donors, financial institutions, and the UN to undertake comprehensive risk assessments to evaluate how specific policies affect conflict. These should "focus on the risks related to recent or on-going conflicts and on potential risks associated with inequality in the distribution of benefits from development."[27] This growing emphasis would inform the development of the United Nations' Post-Conflict Needs Assessment, tools innovated by the OECD-DAC's International Network on Conflict and Fragility, and approaches adopted by the new UN Peacebuilding Commission (see below).

HDR 2005 also shaped international agenda-setting through its advocacy for a "new deal on aid" for conflict-affected countries. Recognizing that starving conflict-prone or post-conflict states of aid is unjustified and bad for both human and global security, the report recommended for donors engaging in war-torn societies: an increased aid effort, greater predictability of aid through long-term financing commitments, and more transparency about the conditions for aid allocations.[28] These policy recommendations were later picked up by the International Dialogue on Peacebuilding and Statebuilding, established by the OECD-DAC and select conflict-affected countries. This culminated in 2011 in the "New Deal for Engagement in Fragile States."

Complementing the Millennium Development Goals (2000–15), the New Deal introduced five peacebuilding and statebuilding goals, as well as guidelines for donor interventions that privileged greater transparency and predictability of aid and the need for "fragility assessments."[29] Several of the conflict-development nexus related themes addressed in *HDR 2005* were also elaborated on in the World Bank's highly influential *World Development Report 2011: Conflict, Security, and Development*.

New global peacebuilding architecture

A third and final area in which the human development school's approach to peacebuilding has shaped the international policy agenda is in connection with the new "Global Peacebuilding Architecture." Once again, *HDR 2005* lent support to efforts within the UN that same year to establish an intergovernmental Peacebuilding Commission (PBC) that would support countries transitioning from war to a durable peace.[30] In short, it recognized the potential of this new body to address a glaring gap in how the UN applied peacebuilding in the 1990s: the lack of institutional capacity for designing policy or providing operational oversight in a way that ensures system-wide coherence by bringing political processes in line with development assistance.[31] Reporting to both the UN Security Council and the General Assembly, the PBC was viewed as an important advance for international peacebuilding by facilitating improved coordination and monitoring of the activities and fund disbursement in a conflict situation from myriad international agency, governmental, civil society, and private sector actors.

Unfortunately, in comprehensive reviews of the Peacebuilding Commission—and the associated Peacebuilding Support Office (PBSO) and Peacebuilding Fund (PBF)—since their creation in late 2005, the Global Peacebuilding Architecture has yet to live up to the full expectations of its founders in terms of achieving a self-sustaining peace in conflict-prone and post-conflict countries.[32] Periodic violent outbursts have occurred in Burundi and Guinea-Bissau, two of the six countries on the PBC agenda, and in a third country, the Central African Republic, violence increased to such a level in late 2013 that the UN requested some 1,600 additional French peacekeepers to assist some 5,000 African Union peacekeepers already in the country. A lack of formal political authority, tools for tracking progress, and the ability to sustain donor financing have limited the Peacebuilding Commission's ability to prevent war recurrence, topics addressed in Chapter 6.

At the same time, through the PBC's country-specific configurations, lessons learned working group, and other innovations in the PBSO and PBF, the commission has facilitated the engagement of new actors and tools in international peacebuilding advocated by *HDR*s since the early 1990s. For instance, reflecting the emphasis placed on nongovernmental organizations as progressive actors for peace, justice, and development in *HDR 1993*, civil society leaders are invited regularly to speak and participate in PBC meetings, the advisory body of the PBC, and the online "community of practice" of the PBSO. *HDR 2005* specifically called for the Global Peacebuilding Architecture to explore new approaches to create "the conditions under which private sector recovery can help to reduce dependence on aid," including public-private partnerships and the use of public finance or public credit guarantees to reduce risk and create incentives for private investment.[33] Again, the PBC and PBSO regularly engage and study how the strengths of the business community can be better employed in peacebuilding, including by leveraging PBF resources to attract long-term financial and technical resources from the private sector in countries recovering from deadly conflict.

New horizon issues

Building on the conceptual and policy advances outlined above stemming from the human development-human security-peacebuilding nexus since the end of the Cold War, several pioneering or "new horizon" issues can be identified. Elaborated on in Chapter 6, three ideas in particular stand out as having the potential to strengthen the efficacy of global institutions in responding to transnational threats to peace and security.

The first is peacebuilding as an organizing framework. The winding down of the wars in Afghanistan and Iraq by the United States, and its mostly Western allies, positions peacebuilding to fill the intellectual void following the disuse of the "global war on terror" organizing principles of counterinsurgency (or COIN) and stabilization operations. Placing a premium on human agency and the need to balance people's security with state security, human development and human security can ensure peacebuilding's long-term utility as an instrument and framework for advancing collective security at global, regional, and national levels.

The second is strengthening the Global Peacebuilding and Conflict Management Architecture. This would involve drawing on lessons from global, regional, national, and sub-national *HDR*s to revisit the mandates, functions, and resources associated with the UN Peacebuilding Commission, Peacebuilding Support Office, and Peacebuilding Fund. Contributing to the 10-year intergovernmental review of the Global Peacebuilding Architecture in 2015, key reform ideas include improving the use of peacebuilding benchmarks and indicators, as well as innovations in mobilizing and sustaining resources from nontraditional donors.

In addition to applying concepts and lessons on human development from fragile and conflict-affected countries, the UN is poised to strengthen the civilian dimensions of a new generation of integrated peace operations better equipped and oriented towards building local, indigenous capacities for conflict management and recovery. Moreover, drawing on its unique contributions to conflict analysis and thinking on human security, the human development school of thought is equipped to help tailor the UN Post-2015 Development Agenda to the special conflict-sensitive needs of the more than 1.5 billion people living in fragile and conflict-affected states.

The third is reforming the UN Security Council to reflect contributions to Human Security. The council was last reformed in 1965, when amendments to UN Charter Article 23 (enlarging its membership from 11 to 15) and Article 27 (on the votes required for procedural and other matters) came into force. To better reflect the contributions of UN member states to advancing human security, participation in Security Council decision-making should be based increasingly on, for example, the level of financial, technical, and political contributions to the civilian and military components of UN peacebuilding and peacekeeping in fragile and conflict-affected countries.

As noted above, each of these "new horizon" issues draws on the interplay between human development, human security, and peacebuilding. Specific characteristics for each, including proposed steps towards their realization, are further elaborated in Chapter 6.

Conclusion

As a result, in large part, of the 40-year Cold War standoff between East and West, distinct political aspects of development—from the choice of national political system to the promotion of political and civil rights—led international bodies, from the UN to the international financial institutions and regional organizations, to steer clear of these "sensitive and often highly charged" issues in the international sphere and, instead, largely dedicate resources and attention to the socio-economic dimensions of development.

Reflecting the changing international policy discourse of the 1990s, the *HDR*s embraced and advanced the previously missing political dimension of development in the research and policy dialogue of global governance actors. Besides favorable conditions, this was inevitable, given that human development remains at its core the expansion of people's choices, whether economic, social, or political in nature.

Combining analytical tools from political scientists, lawyers, and philosophers, in addition to the disciplines of development economics, sociology, and women's studies, thinking on democracy, human rights, and peacebuilding has enriched significantly the concept of human development. Likewise, by giving greater parity to notions of social and economic democracy, the social and economic rights of the individual, and the socio-economic dimensions of peacebuilding, human development has greatly enhanced our understanding and contributed to better applications of these fundamental pillars of both global and national societies today. However, given the four transitions under way in global governance—presented in Chapter 6—key policy and institutional reforms at the global level are necessary to respond to severe risks, prevent backsliding, and ensure the continued relevance of human development in global policy debates.

Notes

1 In 2000, Freedom House classified 120 countries (63 percent of the world total) as democracies. However, the organization argues that by 2013 global freedom had been in decline for its eight consecutive years. See "What's Gone Wrong with Democracy," *The Economist*, 1 March 2014.
2 John Locke, "Popular Basis of Political Authority: John Locke, Second Treatise (First Published in 1698)," in *The Founders' Constitution*, ed. Philip B. Kurland and Ralph Lerner (Chicago, Ill.: University of Chicago Press, 1986), 42.
3 Christopher McMahon, *Authority and Democracy: A General Theory of Government and Management* (Princeton, N.J.: Princeton University Press, 1994), 150.
4 William Connolly, "Introduction: Legitimacy and Modernity," in *Legitimacy and the State*, ed. William Connolly (Oxford: Basil Blackwell, 1984), 10. April Carter reinforces this point when she states, "Despite the fact that conflicts of power and ideology are most intense in the political realm, it is within a political framework that the problem of adapting forms of authority to contemporary conditions has been most successfully resolved: through the evolution of legal tradition, through constitutionalism and representation." April Carter, *Authority and Democracy* (London: Routledge & Kegan Paul, 1979), 71.
5 UNDP, *Human Development Report 2000* (New York: Oxford University Press, 2000), 2 and 31.
6 Ibid., 2.
7 According to *HDR 2002*, "When democratic governments do not respond to the needs of poor people, the public becomes more inclined to support authoritarian or populist leaders who claim that limiting civil liberties and political freedoms will accelerate economic growth and promote social progress and stability." UNDP, *Human Development Report 2002* (New York: Oxford University Press, 2002), 4.
8 UNDP, *Human Development Report 2000*, 2.
9 UNDP, *Human Development Report 2002*, 7.
10 UNDP, *Human Development Report 1994* (New York: Oxford University Press, 1994), 21.
11 UNDP, *Human Development Report 2004* (New York: Oxford University Press, 2004), 6
12 UN General Assembly Resolution 55/2, "United Nations Millennium Declaration," 8 September 2000.
13 UN General Assembly Resolution 62/7, "Support by the United Nations System of the Efforts of Governments to Promote and Consolidate New or Restored Democracies," 8 November 2007. It further reaffirmed in the same resolution "that democracy is a universal value based on the freely expressed will of people to determine their own political, economic, social and cultural systems and their full participation in all aspects of their lives,

Reaffirming also that, while democracies share common features, there is no single model of democracy and that democracy does not belong to any country or region."

14 *The Economist*, "What's gone wrong with democracy," 1 March 2014.
15 UNDP Afghanistan, *Afghanistan Human Development Report 2007: Bridging Modernity and Tradition – Rule of Law and the Search for Justice* (Kabul, Afghanistan: UNDP, 2007), 126–31.
16 UNDP Kosovo, *Kosovo Human Development Report 2004: The Rise of the Citizen – Challenges and Choices* (Pristina, Kosovo: UNDP, 2004).
17 According to the London School of Economics professor Meghnad Desai, the Political Freedom Index came about because the United States was unhappy with its Human Development Index (HDI) ranking and complained that many socialist countries ranked high on the HDI because it ignored freedom. Interview, 4 May 2006.
18 A less politicized argument against these composite indicators that held sway in more academic circles was that human or political freedom could not be reduced to a single number. Both the human and political freedom indices aimed to analyze complex issues with summary responses (either yes or no or a rating of 1 to 10), but as no data and examples were provided for either index, the indices did not allow readers to understand the judgments made behind each index.
19 Edward Mansfield and Jack Snyder, "Democratization and War," *Foreign Affairs* 74, no. 3 (1995): 79–80.
20 World Bank, *World Development Report 2011* (Washington, DC: World Bank, 2011), 1.
21 Andy W. Knight, "Evaluating Recent Trends in Peacebuilding Research," *International Relations of the Asia-Pacific* 3 (2003): 247.
22 Boutros Boutros-Ghali, *An Agenda for Peace*, 2nd edn (New York: United Nations, 1995), 46.
23 UNDP, *Human Development Report 1994*, 23–4.
24 United Nations, "2005 World Summit Outcome," UN doc. A/RES/60/1, 24 October 2005, para. 143, www.un.org/womenwatch/ods/A-RES-60-1-E.pdf.
25 Ibid., paras 138, 139.
26 See www.un.org/humansecurity/, accessed on 30 October 2015.
27 UNDP, *Human Development Report 2005* (Oxford: Oxford University Press, 2005), 169, 180.
28 Ibid., 180.
29 See www.pbsbdialogue.org/en/new-deal/new-deal-principles/, accessed on 30 October 2015.
30 The idea of a Peacebuilding Commission and the related Peacebuilding Support Office and Peacebuilding Fund were proposed originally by the 2004 "Report of the UN Secretary-General's High-Level Panel on Threats, Challenges, and Change."
31 UNDP, *Human Development Report 2005*, 177, 179.
32 For two detailed analyses on the PBC's early years, see Rob Jenkins, *Peacebuilding: From concept to commission* (London: Routledge, 2013); and Mats Berdal, *Building Peace After War* (London: Routledge and the International Institute for Strategic Studies, 2009.)
33 UNDP, *Human Development Report 2005*, 179.

6 Human development in global governance

New frontiers

- **The global governance of ends and means**
- **Four transitions in global governance**
- **Trade**
- **Energy and resources**
- **Sustainable Development Goals and the Post-2015 Development Agenda**
- **Democratic governance and human rights reform recommendations**
- **Peacebuilding reform recommendations**
- **Conclusion**

While achieving significant international recognition since its introduction in 1990, particularly as a perceived counterweight to the Washington Consensus and structural adjustment policies of the World Bank and International Monetary Fund, human development requires renewal and innovation today if it is to continue to shape deliberations and the policies of global institutions. This concluding chapter proposes a rethink of the ends and means of global governance, identifies four new transitions under way in global governance, and considers several "new frontier" issues for which human development analytical tools and policy guidance can continue to make a contribution in the work of global institutions.

The global governance of ends and means

If human development priorities have to be promoted in global institutions, which forums should be chosen? The history of the post-Second World War global order is largely punctuated by the creation of single issue international institutions. So the UN Security Council continues to be the apex body for discussions on international peace and security, even though its legitimacy has been questioned on numerous occasions, whether for the democratic deficit in its membership or its limited ability to constrain great powers from pursuing unilateral actions. The Universal Declaration of Human Rights and its associated covenants had, for more than half a century, the UN High Commissioner for Human Rights to serve as the prime proponent and defender, despite concerns about the balance of treatment of different covenants or the enforcement of basic principles beyond the power of moral suasion. In trade the GATT morphed into the WTO as a forum for rule-making (although developing countries continued to use UNCTAD as a platform to articulate their grievances with the international trading system). The United Nations Framework Convention on Climate Change has remained the main arena for climate negotiations for more than two decades.

In each of the areas discussed in this book the trend is increasingly towards finding multiple points of rule-making, implementation, and enforcement, not to mention legitimacy and participation of various constituencies of interests. The UN Peacebuilding Commission now sits alongside the Security Council while security interventions and peacekeeping missions have also been led by NATO or the African Union. The UN Human Rights Council now adjudicates on the human rights record of member states, while the International Criminal Court has taken the lead for prosecuting crimes against humanity. The WTO's legitimacy was increasingly called into question during the Doha Development Round as the longest running trade negotiation failed to deliver many concrete results (only the December 2013 agreement on trade facilitation has given the WTO some more breathing space to develop broader consensus on other trade issues). Alongside this, major trade negotiations are under way under the Trans-Pacific Partnership and the Transatlantic Trade and Investment Partnership or existing regional forums such as the Asia-Pacific Economic Cooperation (APEC). Hundreds of bilateral investment treaties have been signed to overcome barriers to investment. The glacial progress in climate negotiations has led to calls for finding new venues to pursue climate action or a division of labor for different institutions involved in climate finance, technology development, mitigation action, and interventions for climate adaptation.

Moreover, in certain issue areas, there is still a dearth of institutionalized governance, with organizations and forums either lacking in the scope of their membership or aspects needing attention. The pressures of energy, water, and food can no longer be handled through individual institutions. Yet, for example, there is no global energy regime. Water concerns often manifest through regional politics. Food and agriculture have both humanitarian and industrial scale implications within the same country. To address such under-governed areas, an integrated approach is essential (from village and city levels to the national, regional, and international). New arrangements are needed to tackle the convergence of energy, water, climate, and trade and their further relationship to the broader concept of human development.

The G20 is one forum where developed and large developing countries are discussing some of these concerns. Meanwhile, the BRICS countries have their platforms, offering new forms of international financial assistance but also demanding more voice in global institutions. The danger, however, is that the agendas of international institutions as well as plurilateral groupings of great and emerging powers are shifting rapidly, partly in response to a series of crises (financial, resource-related, ecological, and social) and partly thanks to competition between institutions and shopping for adequately representative forums by member states.

Will human development remain a guiding principle or even have a role in shaping agendas as international institutions evolve and new norms of global governance develop? Institutional evolution is a process but towards what ends? A reform agenda for global institutions draws inspiration from human development priorities as their objectives rather than fitting or limiting human development imperatives within the binds of existing or proposed institutional mechanisms. The choice between ends or means is likely to dictate how countries, their citizens, civil society, transnational networks, and global institutions themselves respond to current deficits in global governance.

Four transitions in global governance[1]

Understanding interconnections between issues and institutions at a global level is a challenging exercise. It is not sufficient to simply analyze governance structures as they exist or their historical origins. There is a need to recognize that global governance is constantly evolving and four on-going transitions are particularly relevant to developing and emerging economies.

First, many emerging economies are gradually shifting from being rule-takers to rule-shapers in global institutions. Whether it is international trade or climate change, developing countries' positions have, historically, often reflected a North-South dynamic. With rapid economic growth yet vast relative and absolute poverty and other social deprivations, countries like India, Brazil, and China now have to straddle the worlds of emerging economies as well as developing countries. The rules that govern international commerce, environment, and access to natural resources have to balance these competing imperatives. The question is how will these countries use their power to shape negotiated outcomes, and on whose behalf?

Second, with a greater voice in negotiations, countries have to identify and articulate their interests not only with regard to specific rules but also on the design of regimes and institutions. Emerging economies are already at the top table of many international regimes, yet many institutions still reflect structures and processes developed for a different era. Regimes are intended to achieve outcomes that states cannot ensure unilaterally. Hence, they have to be designed to deliver functions that add to more than the sum of the interests of their members. With growing power, how would emerging powers influence the redesign of existing international regimes or the creation of new ones that would align with the changing needs of the global system? How will regime design vary for mercantilist issue areas (trade, investment, energy) as means versus those intended to deliver global public goods for human development ends (action on climate change, access to the oceans and their resources, lowering the risk of pandemic diseases)?

Third, there is a shift from singular regimes in specific issue areas to regime complexes, with multiple institutions serving as parallel and nonhierarchical forums for negotiation. Historically, collective security, a founding principle of the UN, has itself been delivered via multiple channels (UN peacekeeping, regional peacekeeping by organizations such as the African Union, formal alliances like NATO, etc.). Other areas like trade and investment have also had competing institutions, such as the GATT/WTO and the UNCTAD. But formal negotiations and rule-setting have mostly occurred in singular institutions (say, the UN Security Council or the WTO's General Council). Now, partially overlapping regimes have gained equal prominence. Rules on trade in energy goods and services are framed in the WTO to some extent, but are far more detailed under the Energy Charter Treaty or in the International Energy Agency (where developing countries are not represented at all).

Or take another example, climate change. Hundreds of billions of dollars of climate finance are expected to flow in the coming decades. It is very likely that only a small fraction of the investments/grants will move through UNFCCC channels. Instead, the governance of multiple sources of financing with dozens of disparate funds is interlinked, often through multilateral banks. How would developing countries emphasize their actions in some forums over others? Does increasing complexity in regimes increase or reduce their freedom of maneuver?

Fourth, developing countries have to recognize that the locus of global governance often follows informal networks and not just formal institutions. The G20 (before it displaced the G7/8), various trade coalitions (the "Quad" countries of the United States, Canada, Japan, and the European Union; the G20 group in the trade regime), or the "BASIC" group—consisting of Brazil, South Africa, India, and China—in climate negotiations are examples that illustrate these trends. Informal networks (whether of government officials or non-governmental experts and organizations) gain prominence and influence when intransigence characterizes formal multilateral negotiations. Networks could develop internal institutional processes to coordinate discussions and actions. But they operate best in small group settings, thus adding to the tension between greater efficiency (promised by few actors) and greater legitimacy (achieved through broader representation). Some larger developing countries are part of many such groupings. But smaller countries are also tied together in recent initiatives, which in turn are led by developed countries (the Climate and Clean Air Coalition is one recent example). With what capacity could these countries manage the trajectory of negotiations as they shift from multilateral forums to informal settings and back again?

In this chapter we outline reform proposals in a range of governance areas. These proposals are best understood as responses to shifts in rule-making capacity and power, the design of new regimes, leveraging the multiplicity of regimes governing any domain, or using networks of state and nonstate actors for human development gains.

Trade

Clean energy and trade rules

Nearly two billion people in the world have no access to modern sources of energy. There remains a vast unmet demand for energy, which is essential for human development. Renewable energy promises to open up opportunities for economic growth for households without electricity, progress in social development in the form of improved education, health, and gender equity outcomes, and offers a credible response to the global challenge of climate change.

Federal and provincial governments across the world have used their authority to promote the development of renewable energy. Yet policies to promote renewable energy often have multiple objectives, in addition to improving outcomes for energy access, such as generating fiscal revenue, developing local industries, creating jobs, and stimulating the economy. Some of these objectives can support renewable energy deployment, and also generate value for the domestic economy across the supply chain of renewable energy research, manufacturing, deployment, and servicing.

But if policies to support renewable energy distort markets, then renewables remain artificially more expensive than they need to be, delaying access to the poor and postponing the day when they can serve as viable substitutes for fossil fuels. Efforts to scale up renewable energy are being obstructed by a range of barriers to sourcing the best technologies from global markets. Subsidies, tariffs, standards, public procurement policies, and local content requirements could slow down or outright hamper possibilities for large-scale use of renewable sources.

Governments are in a dilemma. On one hand, they must recognize that as renewable energy is widely deployed, all countries could benefit, either from improved access to cheaper technologies or newer markets. In fact, a global supply chain for renewable energy products is developing, increasing the likelihood that all countries could benefit from trade in renewable energy products and services. On the other, they need to find ways to promote renewable energy while fulfilling their respective obligations as WTO members. Since 2010, 11 percent of WTO disputes have related, at least in part, to renewable energy. The rise in international trade disputes over renewable energy can serve to hinder efficient renewable energy development and investor confidence across jurisdictions.

Currently, WTO rules do not distinguish between the purposes of government support measures, whether they are subsidies, tax benefits, infrastructure support, or favorable regulations. But the promotion of renewable energy has several benefits, including increasing access to energy (a precondition for human development), and creating a global public good by controlling greenhouse gas emissions. Government support for increasing clean energy access for the poor or deploying clean energy in response to climate change is of a fundamentally different nature to measures that largely seek to expand markets for the sale of clean energy products and services. Treating all types of government support measures in the same manner equates aggressive trade-distorting mercantilist policies with efforts to promote human development.

If the WTO has to play a constructive role in promoting energy access and preventing climate change, its rules have to be modified. To minimize the potential for further disputes in this area, mutually agreed principles must be developed to harmonize international trade and renewable energy policies. Four ideas are:

1 Recognition of the value of policies that support development of high-quality renewable energy manufacturing and do not result in significant distortion of domestic or global markets for renewable energy goods, services, or technologies.
2 The practice of early bilateral consultation between governments on national policies affecting renewable energy trade as one important means of avoiding trade disputes.
3 Circumstances under which two governments would agree to exercise restraint before initiating WTO disputes affecting trade in renewable energy.
4 A Sustainable Energy Trade Agreement, which would distinguish between the purposes of government support and permit exemptions to governments that use a variety of support measures (such as subsidies, feed-in tariffs) to promote energy access or install more renewable energy capacity within their countries while continuing to prohibit or limit the use of trade-distorting government policies aimed at unfairly securing access to export markets.

Energy and resources

The concern over securing predictable access to energy resources dates back at least to the early part of the twentieth century. Despite the continuing quest for energy independence, the reality is that the world's leading economies inhabit a complex world of energy flows complicated by rising demand for energy in the developing world. The resulting multipolarity of energy demand raises concerns about the ability of existing arrangements for energy trade and investment to adapt. Again, while global oil and gas markets are no longer controlled by any single player, the threat of disruption continues.

Some of the largest energy companies are state-owned. The continuing controversy over Gazprom's control over energy supplies into Ukraine—and further into central Europe—underscores the inability of energy markets to act as a sufficient bulwark against state-driven or state-influenced corporate decisions among oil and gas majors. Climate change, as discussed above, has also forced a rethink about the energy mix and the need to develop cleaner energy, whether through more efficient use of fossil fuels or by increasing investments in renewable energy.

Access to energy is closely related to other human development outcomes, whether increase in incomes, better opportunities for education, improvements in household air quality and health outcomes, or reduction in gender disparities. In turn, with rising incomes, the demand for modern energy sources also rises. Households are expected to shift from solid fuels (such as traditional biomass) to liquid fuels (kerosene) to gas (LPG), and eventually electricity. The challenge is that there are no international mechanisms by which energy-consuming and energy-producing nations can balance each other's demands while also ensuring that emerging large energy-demanding countries have security of supply.

Governing supply not just demand

Although much of the focus of climate negotiations and energy modeling scenarios has been on energy demand from emerging economies, it is also important to identify the norms, rules, and institutions that would govern energy supply. For developing countries, with per capita energy consumption a fraction of the levels of developed economies (India's is one-twentieth that of the US, for instance), a human development approach to energy security ought to focus on securing energy supply, not just constraining demand.

A world with multiple poles of energy suppliers, energy demanders, and emerging economies has direct implications for coherence between different international organizations. The countries that are members of the multilateral trade regime do not always overlap with those that are part of producers' cartels. Major energy consumers in the Asia-Pacific region have formed the APEC Energy Working Group. There are new calls for bringing together major suppliers and users under an Energy Stability Board to coordinate emergency actions and give voice to emerging economies. But it is unclear which forums countries will choose to resolve contradictions and disputes.[2]

As a result, there are several deficits in the global governance of energy.[3] These include concerns about the way resource markets operate, the institutions that govern resources, worries about equity in access to the global commons, and the means to mitigate or moderate resource-related conflict. Global energy and food prices have become more volatile in recent years and have also risen to levels not seen since the early 1970s.[4] The success of any country's strategies to secure resources will be, in part, a function of how commodity markets operate and whether large demanders of resources find merit in cooperation rather than in resource nationalism.

Moreover, given the pressures of energy security, climate change, and trade liberalization, it is unlikely that energy trade and investment could be governed through a single institution or regime. In response to the governance deficits, countries will continue to "forum shop," choosing energy governing arrangements that deliver direct benefits, rather than pursue elusive grand bargains. And it is also likely that countries will not surrender their sovereignty over energy issues easily. Rather than aim for an overarching global energy regime or organization, a realistic and pragmatic approach could be to increase the consistency of rules across otherwise fragmented regimes.

A human development approach would focus on increasing access to energy to cover the gaping energy poverty still afflicting more than a billion people on the planet, while reducing the vulnerabilities of the poor and that of future generations. Thus, energy and resource security should be defined as the availability of adequate quantities of critical resources, at prices which are affordable and predictable, with minimum risk of supply disruptions, to ensure sustainability for the environment and future generations. Within such a people-centered framework, the global governance of energy and resources should draw on the following suggestions.

The first is developing optimal infrastructure to deliver energy resources to energy deficit regions. Rather than create competitive infrastructure for, say, gas pipelines or oil refineries, regional networks should be established, which would be more cost-efficient. For instance, China and India could envision a network of pipelines which they could share to tap resources in Russia, Central Asia, and West Asia. As these efforts grow, they could partially mitigate strategic concerns in both countries about the development of energy-related infrastructure in their respective neighborhoods (the Indian Ocean or the South China Sea). Given the political concerns about disruption of energy supplies, investments in pipelines and refineries should have sovereign guarantees from all parties involved.

Second, promoting business models in decentralized energy. There is a strong case for promoting distributed energy infrastructure, through a blend of different renewable energy sources and via smart "micro-grids." This will help to reduce the load on the main grid, offer energy access solutions to those without basic forms of modern energy, and create opportunities for productive uses of renewable energy (such as in small agricultural operations, remote telecom infrastructure, schools and hospitals, etc.). There are hundreds of firms providing decentralized energy services, each experimenting with different business models: selling products like lanterns, installing home systems (solar panels, biogas plants), or developing microgrids. These social entrepreneurs should be supported through a multicountry network to scale alternative business models for distributed generation in Africa, South Asia, Southeast Asia, and remote regions in China that have struggled to get connected to the grid. Moreover, firms in countries with a strong manufacturing base should be encouraged to establish facilities in other markets, to develop local capacity in technology innovation, and counter opposition of dumping and unfair trade practices in renewable energy products.

Third, conducting an internationally coordinated research program, similar to the Green Revolution. These should be directed towards energy storage technologies, and would be critical for both grid stability and off-grid clean energy technologies.

Fourth, establishing an Indo-Pacific Forum (IPF) to promote a regional energy order in the countries where the greatest growth in energy demand is expected over the next few decades. The first task of the IPF would be to increase transparency in energy markets with regular information on oil and gas purchases, long-term contracts, and spot market prices. Secondly, the IPF would facilitate discussions on how each member country's strategic reserves could be used to instill confidence in energy markets to mitigate short-term supply shocks. Thirdly, by offering an open membership platform, the IPF would attract other second tier but rapidly growing energy demanders and collectively press for a reduction in premiums charged on energy supplies to the Indo-Pacific region. Fourthly, members could pool resources to promote technology cooperation in the region. Fifthly, it could discuss protection of key energy supply routes (via land and sea). Sixthly, in a more institutionalized form, it could arbitrate on energy-related disputes and protect overseas investments.

Effective international climate agreements

Reforming global governance for energy, with a view to securing supplies, is one side of the coin. The other is to ensure that any eventual climate agreement is effective in keeping emissions in the atmosphere to a level that would give a reasonable chance of restricting the average surface temperature rise worldwide to under 2°C compared with pre-industrial levels.

For more than two decades this has been a largely fruitless ambition, despite progress in some countries. It is clear that for various reasons certain types of negotiations are not going to succeed. Some of these are discussed below.

One reason is empty promises. The UNFCCC architecture is grounded in certain assumptions—technology leapfrogging, innovative finance, cuts for cash, information-driven response—which have led to a huge rise in expectations. However, nothing serious is on offer to developing countries in terms of their equitable access to the global atmospheric space. Nor has much occurred in delivering new and additional finance. In technology and trade, there remain several barriers to securing new technologies for climate mitigation or adaptation as well as the rising threat of trade disputes. Moreover, information about rising temperatures or warming oceans does not automatically result in adequate response, so a different approach is needed.

Another issue is whether to conduct large-N or small-N discussions. It is clear that negotiations involving 190-plus countries with each one having a veto have not succeeded. But the other extreme of deals between only the United States and China, or BASIC and the US, or at the G20 assume that other countries have nothing to offer or lose (and will be quiet spectators), and that large emitters have no poor people (and so would happily enjoy their seat at the top table)!

(Artificial) coalitions of the willing are another issue. Some emphasis has been laid in recent years on corralling small countries together to apply pressure on large developing countries: for example, the Climate and Clean Air Coalition on short-lived climate pollutants (CCAC). Without China or India, however, there is little expected impact of the CCAC except to harden positions within the negotiations.

Also relevant are values and ethics. For 20 years, climate negotiations have been a "war of values": equitable access to carbon space, intergenerational equity, common but differentiated responsibilities, uncompromising way of life, compensation and polluter pays principle, etc. But their interpretations have been different and the analytics to determine the costs and benefits have been affected by such differing interpretations. Therefore, little common ground has been found to convert the values into enforceable commitments.

An alternative response to climate change would depend on two drivers, both of which are concerned with human development: how soon the future is felt and how soon the costs of the response are passed through to citizens. If too soon, vulnerability (say, sudden shift in energy prices) is exacerbated and the poor are impacted the hardest. But if the response is too late, then the crisis is magnified (with the risk of stranded assets in terms of energy infrastructure, imposition of carbon price in the future, water and food security crises). Neither of the options is politically palatable in any country, rich or poor. If the response has to be just right, an effective climate agreement would offer opportunities to leverage three growing demands: from the poor for access to basic services (and their willingness to pay); from the middle class for better quality of life (energy and water availability, air and water pollution, health impacts, food inflation); and from the rich world and upper middle class for better returns on investments in technology and new business opportunities.

Structuring international climate negotiations and, more importantly, climate action as per the framework of access, efficiency, and externalities would yield a governance process different from that which has failed over two decades.

Climate negotiations structured around access, efficiency, and externalities suggest a different way to aim for small-N deals. Here, the groups would not be divided between developed/developing or Annex I/non-Annex I countries or major economies and the rest. Nor would they be exclusive or exclusionary. Furthermore, this approach would avoid artificial coalitions and also avoid the trap of equity or responsibility (although the final outcomes are intended to be equitable and secure proportional response). Instead, the idea is to bring together countries with clear representational interest in different areas. It would open up the negotiations to a wider suite of interests (including the public and private sector).

The negotiating tracks would no longer center on mitigation, adaptation, technology, and finance. We know that these are often overlapping and contingent on one another. Rather, discussions on "access"-related initiatives would consider impacts on emissions, energy access, adaptability to water stress, various technology options, and sui generis financial models. Discussions on "efficiency" and "externality"-specific initiatives could be structured and evaluated similarly. Efficiency initiatives might include industrial energy efficiency, water efficiency, fuel and water security for power plants, etc. Externality initiatives could include developing chemical alternatives to hydrofluorocarbons, R&D in energy storage technologies, R&D for drought-resistant seeds, or fossil fuel subsidy reform.[5]

Implementation would largely be through private (and in some cases, public sector) participation. This means a confluence of interests of project developers, strategic investors, financiers, insurers, publicly funded risk guarantee schemes, along with R&D and manufacturing supply chains enveloping several economies.

As these would be cross-country projects and investment opportunities, a global carbon market price could also re-emerge. In addition, individual countries that impose a carbon tax need not be necessarily disadvantaged as they would be stimulating cross-border investments rather than simply risking domestic industrial competitiveness. Once the potential is recognized, the threat of border restrictions (via carbon adjustments or trade barriers) would reduce also.

Monitoring and evaluation would be project-specific and determined by the investors and developers, just like in any other sector. However, reporting and verification would be the responsibility of independent agencies (whether national or regional or multilateral).

Would this add up to the effort needed to keep carbon emissions restricted to the trillionth ton of carbon or atmospheric concentration to 450 ppm? It is unlikely, just as all other options being considered so far are also unlikely, to deliver the ideal outcome we are pursuing. But this arrangement has three main benefits over the others:

- It sets out a road map for mitigation and adaptation action at scale rather than merely negotiations over reporting and monitoring.
- It prioritizes action now rather than those potentially driven by unenforceable goals in the future, and thereby reducing the burden on voiceless future generations.
- It overcomes concerns about the voice of small countries in the governance of the climate regime, as small-N groups under such an arrangement would not be exclusive clubs of major economies; rather they would be effective coalitions of parties driven by economic interests in providing access, increasing efficiency, and seeking returns from responding to externalities.

Sustainable Development Goals and the Post-2015 Development Agenda

How should the success of the Human Development Index and related composite indicators be measured against evolving policy priorities? The human development composite indices not only managed to draw attention away from a single crude GDP measure, they were also instrumental in how the international community viewed development priorities in the 1990s and 2000s. Setting the targets for the Millennium Development Goals (MDGs) and tracking their progress subsequently has been a significant undertaking for the UN system; but they are also a consequence of the human development composite indices having been "mainstreamed" into international development priorities. As those priorities evolve, would new indicators be needed?

A case in point is the debate surrounding the Sustainable Development Goals (SDGs) and the UN development agenda beyond 2015. A report of the UN secretary-general in July 2013, *A Life of Dignity for All*, [6] suggested that sustainable development ("enabled by the integration of economic growth, social justice and environmental stewardship") could become the guiding principle and operational standard for the UN. This approach, it was suggested, could continue the work of the MDGs and also end poverty within a generation. Among the various actions being considered are: eradicating poverty in all its forms; empowering women and girls; providing quality education and lifelong learning; improving health; addressing climate change; addressing environmental challenges; promoting inclusive and sustainable growth and decent employment; ending hunger and malnutrition; addressing demographic challenges; enhancing the positive contribution of migrants; meeting the challenges of urbanization; building peace and effective governance via the rule of law and sound institutions; and fostering a renewed global partnership.

As is evident, some of the suggested actions for the SDGs are consistent with and a continuation of the MDGs. Others have been added to account for emerging challenges. But two core problems remain: how would countries determine their priorities among such a long list of intended outcomes; and how would the international community measure progress, if countries exhibited different priorities?

In various debates and statements, developing countries have indicated their concerns about insufficient funds or the means of implementation (such as technology transfer) to achieve another wide slew of development goals. India, for instance, has argued that there might be too much emphasis on the environmental pillar of sustainable development at the expense of the economic and social development aspects of the SDGs. Equally, the G77 and China have prioritized poverty eradication and decent employment as their core concerns. China has resisted mandatory indicators for the "green economy" and has reiterated positions similar to climate change negotiations about common but differentiated responsibilities. Developed countries, such as the United States, have, in turn, drawn attention to technological innovation, investments in education and research, strong intellectual property, conservation of natural ecosystems, and perhaps, most controversially, arguing that the North-South divides between countries are outdated.

The UN General Assembly, in 2015, by securing consensus, partially resolved some of the above concerns and potential contradictions in how countries interpret the wide berth of the SDGs. But the wider the set of concerns and ambitions, the harder it will be to establish common metrics for measuring contributions, performance, and outcomes.

This is where the usefulness of the human development composite indicators might offer insights for what the post-2015 agenda might need. Instead of merely listing targets in individual areas, efforts would be needed to develop new or, at the very least, significantly alter existing human development composite indicators. Some ideas are listed below (as elaborated on in Chapter 3):

- *Resource foot-printing analysis* could combine several dimensions (e.g., water, energy, and carbon footprints) but distinguish the impacts through alternative measurements. The purpose of such analysis would be to draw attention to multiple SDGs (economic development, employment, environmental sustainability, etc.) rather than celebrate successes or shame failures on individual indicators.
- *Green accounting* and accounting for the value of ecosystem services has been suggested for some time. The purpose of new accounting methods would not be to impose one solution on every country, but to offer a toolkit for more informed decision-making.
- A third way in which new indicators would be helpful is in supporting the focus of SDGs on *delivering access to basic services for all*. This focus on reducing inequalities within countries (and not merely measuring aggregate achievements at a country level) would likely correct for a major flaw in how the MDGs were designed; it would also endorse the human development approach by focusing on the entitlements and capabilities of each individual.
- *Integrating political, human rights, and cultural liberty dimensions of human development within the HDI and other human development composite measures.* Learning from earlier attempts to configure a Human Freedom Index and a Political Freedom Index in the 1991 and 1992 *HDR*s, respectively, the conditions are now ripe for a revisit to Mahbub ul Haq's wish to someday fuse political and other dimensions of human development into the Human Development Index. This is a particularly necessary innovation given the expected focus on national and global governance for "delivering on the Post-2015 Development Agenda" following the introduction of the agreed SDGs in September 2015.

The development of composite indicators to align with the SDGs would build on the human development approach in at least three ways: First, by focusing on the individual or the household as the core unit of analysis; second, establishing the core relationships between various development goals; and, as a result, third, by ushering in a "data revolution" and serving as new frameworks for reigniting international development cooperation.

Democratic governance and human rights reform recommendations

As presented in Chapter 5, just as democratic governance and human rights have enriched our understanding and the practice of human development, the inverse is also true; by expanding people's choices and their capabilities for both meeting their basic needs and achieving their highest aspirations, human development has enhanced how democracy and human rights are both comprehended and applied. In particular, two "new frontier" proposals, discussed below, are designed to sustain the relevance of human development in global institutions and international policy debates on strengthening democratic governance and promoting human rights.

Enhancing global governance tools for democratic governance promotion at the national and sub-national levels

Shortly after the 40-year Cold War subsided in 1989, the UN began to invest seriously in the platforms and instruments for strengthening national and sub-national democratic governance. Over the subsequent two and a half decades, it developed new tools—from revitalizing the International Conferences on New and Restored Democracies (which the UN began to support directly in 1994) to increasing the resources, technical assistance, and studies provided through the UN Democracy Fund, UN Department of Political Affairs, UN Development Programme, and UN University—the UN recognized by the early 2000s that it was insufficient to simply help initiate democratic processes; institutions needed to be built and democratic cultures needed to take root, and this took time. New innovations were required to sustain global and national networks for democratic change, including in connection with sub-national geographic units and minority groups that may feel excluded from a country's political, economic, and social progress. Three particular steps that represent the essence of a human development approach to engendering more inclusive and accountable democratic systems of governance are discussed below.

First, the promotion of a democratic culture through the media, civil society, and civil education. Although it may seem a more efficient course of action, it is important that the UN, other global institutions, and the wider donor community not prioritize the needs of the executive branch of government at the expense of other actors. Building and sustaining democratic governance means ensuring proper checks and balances, open competition, and citizens' participation at the earliest stage of democratization. Serious investments, therefore, must be undertaken to prepare effective watchdog bodies in the media and civil society, as well as to inform citizens of their democratic rights and responsibilities. Otherwise, executive-led power politics may soon overwhelm nascent democratic practices and institutions shortly after the executive's election is deemed "free and fair," with adverse consequences for inclusive and transparent governance that is consistent with human development.

Second, keeping democracy affordable and simplifying procedures. Critics of externally supported democratization by global institutions and other international actors complain that it has become far too costly and complex an enterprise. In Afghanistan, for example, the 2004 presidential and 2005 national parliamentary elections alone cost $318 million, exceeding the estimated total government revenue for Afghanistan in 2004 ($269 million). Meanwhile, many countries are also encouraged by foreigners to undertake elections on a frequent basis at various sub-national governance levels, including provinces, districts, and municipalities. Undoubtedly, democracy comes with a price, and countries—particularly when foreign aid declines—must be prepared to sacrifice for their democratic rights. But common sense, careful planning, and some creativity can keep costs down and the management of democratic institutions practical and sensitive to local realities.

Third, accelerating capacity development for elected leaders and civilian administrators in newly established democracies. Democratic authority cannot be embedded in a society simply by signing a legal document that has been accorded international legitimacy, just as peace does not automatically flow from the signing of a peace agreement. Often in transitioning from either a conflict or an authoritarian regime viewed as unjust and non-transparent, newly established democracies must undertake a number of well-resourced capacity development steps to ensure that, over time, the state and its newly elected authorities are firmly established and legitimated vis-à-vis their citizens. Unlike traditional capacity development activities, such as classroom training and foreign study tours, for an initial period of 5–10 years at the start of a new democracy, one-on-one and small group mentoring and coaching is often warranted to build the capacity of elected leaders and mid- to senior-level administrators charged with executing national policy decisions. Besides core technical skills, mentors should emphasize the transfer of "soft" leadership and management skills, such as motivating staff, building a team dynamic, delegating responsibilities, and monitoring and awarding progress. In addition, broader socio-economic foreign assistance to a country in the early phase of democratization should be carefully targeted to buttress and help legitimize newly elected authorities. This implies forgoing UN and broader donor-led approaches that may be expedient in the short term, yet fail to serve the broader goals associated with building durable democratic governance at national and sub-national levels.

Placing economic, social, and cultural rights on par with political and civil rights internationally

In terms of garnering the level of support policy-makers give to the International Covenant on Civil and Political Rights, the International Covenant on Economic, Social, and Cultural Rights—which also came into force in 1976—has come up short since its inception. Admittedly, innovations such as the UN Independent Experts on human rights and extreme poverty (1998) and on the effects of economic reform policies and foreign debt on the full enjoyment of human rights (2000), as well as Special Rapporteur on Economic, Social, and Cultural Rights, and Special Rapporteurs on specific issues such as the human right to education (1998), adequate housing (2000), food (2000), and to safe drinking water and sanitation (2008), have helped to elevate the place of economic, social, and cultural rights on the international agenda. Nevertheless, these efforts remain inadequate. Three key "new frontier" reforms to strike a better balance are discussed below.

First, improving internationally sanctioned independent scrutiny of states' human rights records by strengthening the United Nations' special procedures system, especially with respect to the enforcement of contentious economic, social, and cultural rights within a state.[7] This entails improvements in the appointments process of UN Independent Experts, the resources placed at their disposal, refinement of working methods, increased cooperation by states, more training opportunities, and enhanced coordination with relevant UN system bodies and agencies. However, the most important reform involves measures to strengthen the follow-up and implementation of UN Independent Experts, such as requiring mid-term status reports (a practice already begun by several states), encouraging treaty bodies to do more to consult and build on the relevant UN Independent Experts' recommendations, and encouraging national human rights institutions, ombudspersons, and parliamentary bodies to play important follow-up roles as independent facilitators and advocates.[8]

Second, the UN High Commissioner for Human Rights and the UN Human Rights Council should offer similar resources, status, and high-level political attention to economic, social, and cultural rights as they give to political and civil rights in global governance, including through focused research, special policy reports, and frequent high-level policy dialogues dedicated to economic, social, and cultural rights.

Third, building on the path-breaking *Human Development Report 2000: Human Rights and Human Development* (Chapter 5) on "Using indicators for human rights accountability." This is to strengthen the role of civil society actors in monitoring state compliance with their human rights commitments, undertaking advocacy campaigns to press for reforms, and keeping UN Independent Experts informed of progress and setbacks in the enforcement of human rights, including often neglected and misunderstood economic, social, and cultural rights. In short, statistics can help to raise important questions behind general arguments and help to reveal broader social challenges in a society.[9] When applied extensively and effectively to a society with a problematic human rights record, indicators also can be useful in identifying the unintended consequences of laws, policies, and practices; they can prompt preventive action, by giving early warning of potential violations; and especially in connection with contentious economic, social, and cultural rights, they can enhance social consensus on difficult tradeoffs to be made in the face of resource constraints.[10]

Peacebuilding reform recommendations

Similar to democratic governance and human rights, peacebuilding has also affected how human development is understood and applied, particularly towards the role of global institutions in responding to state fragility and conflict. A successor concept to human security—pioneered in the *Human Development Report 1994*—peacebuilding rests at the intersection of conflict, security, and development. With the advent of the New Peacebuilding Architecture in 2005, consisting of the UN Peacebuilding Commission, the Peacebuilding Support Office, and the Peacebuilding Fund, peacebuilding has risen to prominence over the past decade in some international policy circles. To continue to expand its global reach, there are three primary new frontier reform initiatives, which are discussed below.

Peacebuilding as an organizing framework

Given the scale, scope, and attempted reach and ambitions of the operations in Afghanistan and Iraq since the turn of the century, stabilization and counterinsurgency (or COIN) operations largely enabled the military's objectives to trump the diplomatic and development objectives of civilians. As the United States and its Western allies wind down these wars, peacebuilding is poised to fill the intellectual void of these "global war on terror" organizing principles. Contrary to common definitions for stabilization and COIN, distinguishing features of peacebuilding as a new organizing framework for global institutions in fragile and conflict-affected states—consistent with its use in the *HDR*s—should include:

- Fundamentally civilian-led efforts to build host nation institutional and human capacities for managing and addressing the sources of violent conflict.
- Second, a commitment to UN-led multilateral approaches to building local and sustainable capacities for peace.
- Finally, the need to actively source talent from beyond Western donor countries, especially as Diasporas and experts from the Global South often can bring scarce language skills and cultural sensitivity, in addition to technical expertise, to a peacebuilding mission.

In short, given that human development and the forerunner to peacebuilding—human security—place a premium on human agency and the need to balance people's security with state security, both human development and human security can help to ensure peacebuilding's long-term utility as an instrument and framework for advancing collective security at global, regional, and national levels.

Strengthening the global peacebuilding and conflict management architecture

Reforms under this heading can be grouped in three areas. First, is the future of the new peacebuilding architecture. In 2015, a 10-year intergovernmental review of the UN Peacebuilding Commission (PBC), Peacebuilding Support Office (PBSO), and Peacebuilding Fund (PBF) is planned. Key ideas for strengthening this new framework for UN peacebuilding support include: (a) expanding the Commission's reach by returning to the original expectation, back in 2005, of inviting each year four-to-five conflict-affected countries to benefit from the work of the PBC; (b) measuring and facilitating the Commission's progress toward consolidating durable peace through concrete, time-bound, and measurable benchmarks; and (c) redoubling efforts within the PBC to marshal, align, and sustain donor resources in support of integrated peacebuilding strategy objectives and commitments.

Second, strengthening the civilian dimension of integrated UN peace operations. Over the past decade, with the advent of new offices such as the UK's Stabilization Unit, Canada's START, and the US State Department's Bureau of Conflict and Stabilization, the international community has come a long way in nurturing and retaining human talent for peacebuilding. However, despite repeated calls from the G8 and other international forums, as well as important innovations such as the UN's CAP-MATCH initiative, much more can be done to foster civilian capabilities for Integrated UN Peace Operations from the Global South, which maintains a talent-base and skill-sets most in demand for building local, indigenous capacities for conflict management and recovery. Here, the UN Peacebuilding Commission can serve as a champion and the PBSO can gather lessons from the various initiatives to train UN peacekeepers and police units, as well as UNDP's Capacity-Building Facilities, which source highly specialized coaches and mentors from the neighboring countries to a conflict zone.

Third, developing conflict-sensitive SDGs. Just as human development priorities in the 1990s served as forerunners to the MDGs (2000–2015), the *HDR*s are informing the definition and associated time-bound targets for the successor SDGs (2015–2030), as elaborated through the UN Post-2015 Development Agenda. The human development school of thought is also equipped, drawing on its unique contributions to conflict analysis and thinking on human security, to help tailor the Post-2015 Development Agenda SDGs to the more than 1.5 billion inhabitants of weak and war-affected states. Such efforts should build directly on A New Deal for Engagement in Fragile States, which has built a diverse international consensus around five major peacebuilding and statebuilding goals emanating from a three-year International Dialogue on Peacebuilding and Statebuilding (2009–2011).[11]

Reforming the UN Security Council to reflect contributions to human security

The last time the UN Security Council was reformed was 50 years ago, in 1965, some 20 years after the founding of the UN. At the time, amendments to Article 23 (enlarging the Security Council's membership from 11 to 115) and Article 27 (on the votes required for procedural and other matters in the Security Council) of the UN Charter came into force. From the 2002 *Human Development Report* and 2004 *High-Level Panel on Threats, Challenges, and Change* to more recent studies by scholars and policy analysts,[12] much ink has been spilled on the need to base Security Council decision-making increasingly on, for example, the level of financial, technical, and political contributions to the civilian and military components of UN peacebuilding and peacekeeping in fragile and conflict-affected countries. This would help to ensure that those UN member states most capable of shaping human security outcomes are included in the Security Council's membership. A range of models have been proposed in this regard, normally expanding the Security Council by 5–10 new permanent members consisting of large countries from the Global North and South (e.g., Brazil, Germany, India, and Japan). In lieu of a more far-reaching reform—and hence amendment to the UN Charter, another novel idea would be simply to remove from Article 23, section II of the UN Charter the line: "A retiring member shall not be eligible for immediate re-election." This would allow for the possibility, at least, that large countries from both the Global North and South could remain as sustained contributors to UN Security Council decision-making for an extended period.

Conclusion

This book sought to examine the idea of human development and its influence, since the end of the Cold War, in transforming our understanding of human progress in institutions of global governance. Defined as "a process of enlarging people's choices" to improve the human condition, human development represents both an intellectual and policy breakthrough in global governance, given its success in reminding world leaders, scholars, and civil society at large of the ultimate purpose of development: to treat all people—present and future generations—as ends. When healthy, educated, well-nourished, and empowered, people are also the chief means of development.

Through a review of the concepts, analysis, and policy and institutional reform recommendations of UNDP's flagship *HDR* series since its inception in 1990, human development has contributed to a shift in academic and policy arenas away from national income accounting to people, their well-being, and the human capabilities to expand their well-being. This was reflected in the outcome documents of UN world summits in the early 1990s, to the advent of the United Nations' MDGs (2000–15), to the announcement in New York in 2015 of the SDGs (2015–2030) at the heart of the UN's Post-2015 Development Agenda.

In sum, the concept of human development and its skillful advocacy through the *HDR*s are shown to have shaped multilateral policy discourse over the past 25 years in global institutions in at least three distinct ways: (i) shifting and expanding what is commonly perceived as the goal of development today; (ii) innovating new means for gauging human and national progress; and (iii) introducing many practical recommendations for change in international public policy. They garnered particular international recognition in the early 1990s as a critical response and alternative to the 1980s structural adjustment policies of the international financial institutions, and to a preoccupation by wealthy donor nations with GNP, per-capita income, and other national income accounting tools.

At the same time, a multiplicity of factors, ranging from the international financial crisis of 2008 to terrorism and other forms of extremism, resource scarcity and climate change, have converged today to place at risk both a continued commitment to human development priorities and the tangible gains achieved in recent decades. As this book argues, including through the presentation of a diverse reform agenda in this chapter, the reinvigoration of international policy debates about human development is vital to rejuvenate and sustain existing global institutions, such as the UN, World Bank, and World Trade Organization as well as new and emerging institutions. This is especially the case with regards to "new frontier issues," such as energy, the environment, peacebuilding, and democratization, at all levels of governance. And in doing so, the renewal of global governance can help, in turn, to enrich and expand both thinking about and the application of human development to understanding and responding to real-world problems.

From its origins in the classical economics of Adam Smith, David Ricardo, Sir William Petty, and John Stuart Mill to the writings of philosophers from both the East (Kautilya) and West (Plato and Aristotle), human development has reintroduced a political economy approach to an examination of today's global challenges. By emphasizing principles of care, fairness, equity, and good democratic governance, it has further injected the ethic of social, economic, and political justice into the deliberations, programs, and culture of global institutions. Indeed, human development and global institutions maintain a symbiotic relationship, and if global institutions are to remain relevant, let alone thrive, in the twenty-first century and beyond, international policy-makers and the citizens who bestow authority on them would be wise to employ the analytical tools and policy guidance of the proponents of human development.

Notes

1 This discussion draws on Arunabha Ghosh, Arundhati Ghose, Suman Bery, C. Uday Bhaskar, Tarun Das, Nitin Desai, Anwarul Hoda, Kiran Karnik, Srinivasapuram Krishnaswamy, Radha Kumar, and Shyam Saran, "Understanding Complexity, Anticipating Change," Report of the Working Group on India and Global Governance (New Delhi: Council on Energy, Environment and Water, 2011).

2 Arunabha Ghosh, "Seeking Coherence in Complexity? The Governance of Energy by Trade and Investment Institutions," *Global Policy* 2 (2011): 109.

3 Arunabha Ghosh, "Three Mantras for India's Resource Security," *Seminar* 658 (2014): 47.

4 Arunabha Ghosh and David Steven "India's Energy, Food, and Water Security: International Cooperation for Domestic Capacity," in *Shaping the Emerging World: India and the Multilateral Order*, ed. Waheguru Pal Singh Sidhu, Pratap Bhanu Mehta, and Bruce Jones (Washington, DC: Brookings Institution Press, 2013).

5 For the design of new and effective climate technology partnerships, see Arunabha Ghosh, and Sudatta Ray "Fixing Climate Governance Through Effective Technology Partnerships," *CIGI Fixing Climate Governance Series – Paper No. 3*, November, (Waterloo, ON: Centre for International Governance Innovation, 2015). www.cigionline.org/publications/fixing-climate-governance-through-effective-technology-partnerships.

6 Report of the Secretary-General, "A Life of Dignity for All: Accelerating Progress Towards the Millennium Development Goals and Advancing the United Nations Development Agenda beyond 2015," UN doc. A/68/202, 26 July 2013.

7 For a detailed assessment and concrete set of recommendations to improve the UN's special procedures system, see Ted Piccone, *Catalysts for Change: How the UN's Independent Experts Promote Human Rights* (Washington, DC: Brookings Institution Press, 2012).

8 Ibid., 123–4.

9 UNDP, *Human Development Report 2000* (Oxford: Oxford University Press, 2000), 90.

10 Ibid., 89.

11 For further details on A New Deal for Engagement in Fragile States, see www.pbsbdialogue.org/en/new-deal/new-deal-principles/, accessed on 30 October 2015.

12 For example, see Richard Gowan and Nora Gordon, "Pathways to Security Council Reform," New York University Center on International Cooperation, May 2014.

Select bibliography

Human development

Sudhir Anand and Amartya Sen, "Human Development and Economic Sustainability," *World Development* 28, no. 12 (2000): 2029–2049. A seminal paper, which argued for sustainable human development, by applying an ethic of "universalism" of claims within and between generations.

Sakiko Fukuda-Parr and A.K. Shiva Kumar, eds, *Readings in Human Development: Concepts, Measures and Policies for a Development Paradigm* (New Delhi: Oxford University Press, 2003). This volume compiles many of the leading articles and *Human Development Report* chapters on the concept, measurement, and policy impact of human development in the early years of the *HDR*s.

Khadija Haq and Richard Ponzio, eds, *Pioneering the Human Development Revolution: An Intellectual Biography of Mahbub ul Haq* (New Delhi: Oxford University Press, 2008). The book is a collection of writings from many of Mahbub ul Haq's closest collaborators highlighting the evolution of Dr. Haq's thinking on human development (and its forerunners) at key intervals of his career.

Mahbub ul Haq, *Reflections on Human Development* (New York: Oxford University Press, 1995). This monograph compiles key writings of Mahbub ul Haq that fed into the earliest *HDR*s (1990–1995).

Richard Jolly, Louis Emmerij, and Thomas Weiss, *UN Ideas that Changed the World* (Bloomington: Indiana University Press, 2009). This volume includes an account of how UN staff conceived of and shepherded the development of the influential concept of human development.

Desmond McNeill, "Human Development: The Power of the Idea," *Journal of Human Development* 8, no. 1 (2007). This article traces the evolution of the concept of human development and its intellectual and policy influence since the inception of the *HDR*s in 1990.

UNDP, *Human Development Report 1990* (New York: Oxford University Press, 1990). This inaugural *HDR* introduces the concept of human development and its flagship measurement tool, the Hum Dev Index.

UNDP, *Human Development Report 1992* (New York: Oxford University Press, 1992). This *HDR* argues that restricted global markets in key agriculture and textile sectors, coupled with barriers to immigration, costs developing countries far more than they receive in foreign aid.

UNDP, *Human Development Report 2010* (New York: UN Development Programme, 2010). This 20th anniversary edition of the *HDR* looks back at the past several decades and identifies often surprising trends and patterns with important lessons for the future of human development.

Sustainable development

The World Commission on Environment and Development, *Our Common Future* (Oxford: Oxford University Press, 1987). The report of the Brundtland Commission squarely put sustainability and development concerns within a single framework.

United Nations, "Resolution adopted by the General Assembly on 27 July 2012: The future we want," General Assembly, doc. A/RES/66/288 (11 September 2012). www.un.org/ga/search/view_doc.asp?symbol=A/RES/66/288&Lang=E. This was the outcome document of the Rio+20 Summit, which initiated the process of drafting the Sustainable Development Goals.

UNDP, *Human Development Report 1998* (New York: Oxford University Press, 1998). This *HDR* focused on changing consumption patterns and their impact on sustainability for future generations.

Global governance and global institutions

Commission on Global Security, Justice & Governance, *Confronting the Crisis of Global Governance* (The Hague and Washington, DC: The Hague Institute for Global Justice and the Stimson Center, 2015). The report of the Albright/Gambari Commission offers a vision for just security and presents an action plan for innovating global governance on key issues at the intersection of security and justice.

Arunabha Ghosh, Arundhati Ghose, Suman Bery, C. Uday Bhaskar, Tarun Das, Nitin Desai, Anwarul Hoda, Kiran Karnik, Srinivasapuram Krishnaswamy, Radha Kumar, and Shyam Saran, "Understanding Complexity, Anticipating Change," *Report of the Working Group on India and Global Governance* (New Delhi: Council on Energy, Environment and Water, 2011). This report offers Indian perspectives on the complex landscape of global governance, shifting rules and norms, the design of international institutions, overlapping (and, at times, contradicting) regimes and mandates, and the influence of emerging networks (formal and informal).

Craig Murphy, *The United Nations Development Programme: A Better Way?* (Cambridge: Cambridge University Press, 2006). This book traces the evolution of the UN Development Programme since its inception in the 1960s, including UNDP's intellectual and policy contributions through such innovative studies as the annual global *Human Development Report*.

Thomas G. Weiss, *Global Governance: Why? What? Whither?* (Cambridge: Polity Press, 2013). This probing yet accessible book examines "global governance" or the sum of the informal and formal values, norms, procedures, and institutions that help states, intergovernmental organizations, civil society, and transnational corporations identify, understand, and address transboundary problems.

Key human development-related sectors covered in this book

Democracy and human rights

Richard Ponzio, *Democratic Peacebuilding: Aiding Afghanistan and other Fragile States* (Oxford: Oxford University Press, 2011). This book examines the evolution of international peacebuilding since the end of the Cold War, identifying the factors that limit the progress of international actors to institutionalize democratic authority and the rule of law in war-shattered societies.

UNDP, *Human Development Report 2000* (New York: Oxford University Press, 2000). This *HDR* pushes the conceptual, analytical, and measurement boundaries on the mutually reinforcing relationship between human rights and human development.

UNDP, *Human Development Report 2002* (New York: Oxford University Press, 2002). This *HDR* considers how human development benefits from and affects the practice and culture of democracy, including in fragile and conflict-affected environments.

UNDP, *Human Development Report 2004* (New York: Oxford University Press, 2004). This *HDR* rethinks the relationships between democracy, human rights, and human development by arguing in favor of protecting and strengthening cultural liberties.

UNDP Afghanistan, *Afghanistan Human Development Report 2007: Bridging Modernity and Tradition – Rule of Law and the Search for Justice* (Kabul, Afghanistan: UNDP, 2007). This national *HDR* argues for the need for a "hybrid model of justice" that combines elements of modern and traditional approaches to justice for a transitional period to improve access to a fair system of justice for all Afghans.

UNDP Kosovo, *Kosovo Human Development Report 2004: The Rise of the Citizen – Challenges and Choices* (Pristina, Kosovo: UNDP, 2004). This national *HDR* examines the policy challenges and choices to expand human development and democratic practice for all Kosovans.

Energy and climate change

David King, Daniel Schrag, Zhou Dadi, Qi Ye, and Arunabha Ghosh, *Climate Change: A Risk Assessment*, (London: UK Foreign and Commonwealth Office, July 2015). The first comprehensive risk assessment of climate change with a focus on the direct and indirect impacts on food, water, public health, drought, and flooding, as well as migration, state fragility and human security.

Nicholas Stern, *The Economics of Climate Change. The Stern Review* (Cambridge: Cambridge University Press, 2006). The seminal document, which analyzed the costs of acting on climate change sooner rather than later and used a low discount rate to value the life chances of future generations expected to be affected by climate change.

UNDP, *Human Development Report 2007/2008: Fighting Climate Change: Human Solidarity in a Divided World*, (New York: Palgrave Macmillan, 2007). The hard-hitting *HDR*, which highlighted the massive inequalities in access to energy and to the global carbon space, the resulting impact on human development and exposed the unmet promises of clean technology and finance for the world's poorest countries.

Peacebuilding

Mats Berdal, *Building Peace After War* (London: Routledge and the International Institute for Strategic Studies, 2009). This book provides a concise overview of the concept of peacebuilding, the new UN Peacebuilding Architecture, and strategies for achieving sustainable peace.

Boutros Boutros-Ghali, *An Agenda for Peace*, 2nd edn (New York: United Nations, 1995). This path-breaking paper by the UN Secretary-General pushes the boundaries of UN thinking on conflict prevention, peacemaking, peacekeeping, and post-conflict peacebuilding.

International Dialogue on Peacebuilding and Statebuilding, "New Deal for Engagement in Fragile States", www.newdeal4peace.org/wp-content/uploads/2013/01/new-deal-for-engagement-in-fragile-states-en.pdf, 2011. The "New Deal" commits participating donor and partner countries to advancing the five peacebuilding and statebuilding goals of legitimate politics, security, justice, economic foundations, and revenues and services.

Rob Jenkins, *Peacebuilding: From Concept to Commission* (London: Routledge, 2013). This volume provides a detailed analysis on the UN Peacebuilding Commission's early years.

UNDP, *Human Development Report 1994* (New York: Oxford University Press, 1994). This early *HDR* conceptualized sustainable human development and introduced the idea of human security.

UNDP, *Human Development Report 2005* (Oxford: Oxford University Press, 2005). This *HDR* includes an in-depth analysis of how conflict affects human development and how human development strategies can help to address the sources of violent conflict.

United Nations, "Report of the UN Secretary-General's High-Level Panel on Threats, Challenges, and Change", 2004. This seminal report set the stage for the 2005 UN Summit and introduced, among other proposals, the idea of a UN Peacebuilding Commission.

World Bank, *World Development Report 2011* (Washington, DC: World Bank, 2011). On the theme of Conflict, Security, and Development, this *WDR* documents the challenges organized violence poses to the advancement of less developed countries.

Trade

UNDP, *Making Global Trade Work for People* (London: Earthscan, 2003). This volume emphasized the need to promote trade with a human face.

UNDP, *Human Development Report 2005: International Cooperation at a Crossroads: Aid, Trade and Security in an Unequal World* (New York: Oxford university Press, 2005). This combination of several core concerns of the international development agenda in this *HDR* helped to shape the discussion at the UN World Summit in 2005 and in other international forums.

Water

UNDP, *Human Development Report 2006: Beyond Scarcity: Power, Poverty and the Global Water Crisis*, (New York: Palgrave Macmillan, 2006). The report was published in the first year of the International Decade for Action "Water for Life" and was instrumental in framing access to water as a consequence of inequality and power politics.

Index

Routledge Global Institutions Series

115 Human Development and Global Institutions (2016)
Evolution, impact, reform
by Richard Ponzio (The Hague Institute for Global Justice) and Arunabha Ghosh (Council on Energy, Environment and Water)

114 NGOs and Global Trade (2016)
Non-State Voices in EU Trade Policymaking
by Erin Hannah (University of Western Ontario)

113 Brazil as a Rising Power (2016)
Intervention Norms and the Contestation of Global Order
edited by Kai Michael Kenkel (IRI/PUC-Rio) and Philip Cunliffe (University of Kent)

112 The United Nations as a Knowledge System (2016)
by Nanette Svenson (Tulane University)

111 Summits and Regional Governance (2016)
The Americas in comparative perspective
edited by Gordon Mace (Université Laval), Jean-Philippe Thérien (Université de Montréal), Diana Tussie (Facultad Latinoamericana de Ciencias Sociales), and Olivier Dabène (Sciences Po)

110 Global Consumer Organizations (2015)
by Karsten Ronit (University of Copenhagen)

109 Expert Knowledge in Global Trade (2015)
edited by Erin Hannah (University of Western Ontario), James Scott (King's College London), and Silke Trommer (University of Helsinki)

108 World Trade Organization (2nd edition, 2015)
Law, economics, and politics
by Bernard M. Hoekman (European University Institute) and Petros C. Mavroidis (European University Institute)

107 Women and Girls Rising (2015)
Progress and resistance around the world
by Ellen Chesler (Roosevelt Institute) and Theresa McGovern (Columbia University)

106 The North Atlantic Treaty Organization (2nd edition, 2015)
by Julian Lindley-French (National Defense University)

105 The African Union (2nd edition, 2015)
by Samuel M. Makinda (Murdoch University), F. Wafula Okumu (The Borders Institute), and David Mickler (University of Western Australia)

104 Governing Climate Change (2nd edition, 2015)
by Harriet Bulkeley (Durham University) and Peter Newell (University of Sussex)

103 The Organization of Islamic Cooperation (2015)
Politics, problems, and potential
by Turan Kayaoglu (University of Washington, Tacoma)

102 Contemporary Human Rights Ideas (2nd edition, 2015)
by Bertrand G. Ramcharan

101 The Politics of International Organizations (2015)
Views from insiders
edited by Patrick Weller (Griffith University) and Xu Yi-chong (Griffith University)

100 Global Poverty (2nd edition, 2015)
Global governance and poor people in the post-2015 era
by David Hulme (University of Manchester)

99 Global Corporations in Global Governance (2015)
by Christopher May (Lancaster University)

98 The United Nations Centre on Transnational Corporations (2015)
Corporate conduct and the public interest
by Khalil Hamdani and Lorraine Ruffing

97 The Challenges of Constructing Legitimacy in Peacebuilding (2015)
Afghanistan, Iraq, Sierra Leone, and East Timor
by Daisaku Higashi (University of Tokyo)

96 The European Union and Environmental Governance (2015)
by Henrik Selin (Boston University) and
Stacy D. VanDeveer (University of New Hampshire)

95 Rising Powers, Global Governance, and Global Ethics (2015)
edited by Jamie Gaskarth (Plymouth University)

94 Wartime Origins and the Future United Nations (2015)
edited by Dan Plesch (SOAS, University of London) and
Thomas G. Weiss (CUNY Graduate Center)

93 International Judicial Institutions (2nd edition, 2015)
The architecture of international justice at home and abroad
by Richard J. Goldstone (Retired Justice of the Constitutional Court of South Africa) and Adam M. Smith (International Lawyer, Washington, DC)

92 The NGO Challenge for International Relations Theory (2014)
edited by William E. DeMars (Wofford College) and
Dennis Dijkzeul (Ruhr University Bochum)

91 21st Century Democracy Promotion in the Americas (2014)
Standing up for the polity
by Jorge Heine (Wilfrid Laurier University) and
Brigitte Weiffen (University of Konstanz)

90 BRICS and Coexistence (2014)
An alternative vision of world order
edited by Cedric de Coning (Norwegian Institute of International Affairs), Thomas Mandrup (Royal Danish Defence College), and Liselotte Odgaard (Royal Danish Defence College)

89 IBSA (2014)
The rise of the Global South?
by Oliver Stuenkel (Getulio Vargas Foundation)

88 Making Global Institutions Work (2014)
edited by Kate Brennan

87 Post-2015 UN Development (2014)
Making change happen
edited by Stephen Browne (FUNDS Project) and Thomas G. Weiss (CUNY Graduate Center)

86 Who Participates in Global Governance? (2014)
States, bureaucracies, and NGOs in the United Nations
by Molly Ruhlman (Towson University)

85 The Security Council as Global Legislator (2014)
edited by Vesselin Popovski (United Nations University) and Trudy Fraser (United Nations University)

84 UNICEF (2014)
Global governance that works
by Richard Jolly (University of Sussex)

83 The Society for Worldwide Interbank Financial Telecommunication (SWIFT) (2014)
Cooperative governance for network innovation, standards, and community
by Susan V. Scott (London School of Economics and Political Science) and Markos Zachariadis (University of Cambridge)

82 The International Politics of Human Rights (2014)
Rallying to the R2P cause?
edited by Monica Serrano (Colegio de Mexico) and Thomas G. Weiss (The CUNY Graduate Center)

81 Private Foundations and Development Partnerships (2014)
American philanthropy and global development agendas
by Michael Moran (Swinburne University of Technology)

80 Nongovernmental Development Organizations and the Poverty Reduction Agenda (2014)
The moral crusaders
by Jonathan J. Makuwira (Royal Melbourne Institute of Technology University)

79 Corporate Social Responsibility (2014)
The role of business in sustainable development
by Oliver F. Williams (University of Notre Dame)

78 Reducing Armed Violence with NGO Governance (2014)
edited by Rodney Bruce Hall (Oxford University)

77 Transformations in Trade Politics (2014)
Participatory trade politics in West Africa
by Silke Trommer (Murdoch University)

76 Rules, Politics, and the International Criminal Court (2013)
by Yvonne M. Dutton (Indiana University)

75 Global Institutions of Religion (2013)
Ancient movers, modern shakers
by Katherine Marshall (Georgetown University)

74 Crisis of Global Sustainability (2013)
by Tapio Kanninen

73 The Group of Twenty (G20) (2013)
by Andrew F. Cooper (University of Waterloo) and Ramesh Thakur (Australian National University)

72 Peacebuilding (2013)
From concept to commission
by Rob Jenkins (Hunter College, CUNY)

71 Human Rights and Humanitarian Norms, Strategic Framing, and Intervention (2013)
Lessons for the Responsibility to Protect
by Melissa Labonte (Fordham University)

70 Feminist Strategies in International Governance (2013)
edited by Gülay Caglar (Humboldt University, Berlin), Elisabeth Prügl (the Graduate Institute of International and Development Studies, Geneva), and Susanne Zwingel (the State University of New York, Potsdam)

69 The Migration Industry and the Commercialization of International Migration (2013)
edited by Thomas Gammeltoft-Hansen (Danish Institute for International Studies) and Ninna Nyberg Sørensen (Danish Institute for International Studies)

68 Integrating Africa (2013)
Decolonization's legacies, sovereignty, and the African Union
by Martin Welz (University of Konstanz)

67 Trade, Poverty, Development (2013)
Getting beyond the WTO's Doha deadlock
edited by Rorden Wilkinson (University of Manchester) and James Scott (University of Manchester)

66 The United Nations Industrial Development Organization (UNIDO) (2012)
Industrial solutions for a sustainable future
by Stephen Browne (FUNDS Project)

65 The Millennium Development Goals and Beyond (2012)
Global development after 2015
edited by Rorden Wilkinson (University of Manchester) and David Hulme (University of Manchester)

64 International Organizations as Self-Directed Actors (2012)
A framework for analysis
edited by Joel E. Oestreich (Drexel University)

63 Maritime Piracy (2012)
by Robert Haywood (One Earth Future Foundation) and Roberta Spivak (One Earth Future Foundation)

62 United Nations High Commissioner for Refugees (UNHCR) (2nd edition, 2012)
by Gil Loescher (University of Oxford), Alexander Betts (University of Oxford), and James Milner (University of Toronto)

61 International Law, International Relations, and Global Governance (2012)
by Charlotte Ku (University of Illinois)

60 Global Health Governance (2012)
by Sophie Harman (City University, London)

59 The Council of Europe (2012)
by Martyn Bond (University of London)

58 The Security Governance of Regional Organizations (2011)
edited by Emil J. Kirchner (University of Essex) and Roberto Domínguez (Suffolk University)

57 The United Nations Development Programme and System (2011)
by Stephen Browne (FUNDS Project)

56 The South Asian Association for Regional Cooperation (2011)
An emerging collaboration architecture
by Lawrence Sáez (University of London)

55 The UN Human Rights Council (2011)
by Bertrand G. Ramcharan (Geneva Graduate Institute of International and Development Studies)

54 Responsibility to Protect (2011)
Cultural perspectives in the Global South
edited by Rama Mani (University of Oxford) and Thomas G. Weiss (The CUNY Graduate Center)

53 The International Trade Centre (2011)
Promoting exports for development
by Stephen Browne (FUNDS Project) and Sam Laird (University of Nottingham)

52 The Idea of World Government (2011)
From ancient times to the twenty-first century
by James A. Yunker (Western Illinois University)

51 Humanitarianism Contested (2011)
Where angels fear to tread
by Michael Barnett (George Washington University) and Thomas G. Weiss (The CUNY Graduate Center)

50 The Organization of American States (2011)
Global governance away from the media
by Monica Herz (Catholic University, Rio de Janeiro)

49 Non-Governmental Organizations in World Politics (2011)
The construction of global governance
by Peter Willetts (City University, London)

48 The Forum on China-Africa Cooperation (FOCAC) (2011)
by Ian Taylor (University of St. Andrews)

47 Global Think Tanks (2011)
Policy networks and governance
by James G. McGann (University of Pennsylvania) with Richard Sabatini

46 United Nations Educational, Scientific and Cultural Organization (UNESCO) (2011)
Creating norms for a complex world
by J.P. Singh (Georgetown University)

45 The International Labour Organization (2011)
Coming in from the cold
by Steve Hughes (Newcastle University) and Nigel Haworth (University of Auckland)

44 Global Poverty (2010)
How global governance is failing the poor
by David Hulme (University of Manchester)

43 Global Governance, Poverty, and Inequality (2010)
edited by Jennifer Clapp (University of Waterloo) and Rorden Wilkinson (University of Manchester)

42 Multilateral Counter-Terrorism (2010)
The global politics of cooperation and contestation
by Peter Romaniuk (John Jay College of Criminal Justice, CUNY)

41 Governing Climate Change (2010)
by Peter Newell (University of East Anglia) and
Harriet A. Bulkeley (Durham University)

40 The UN Secretary-General and Secretariat (2nd edition, 2010)
by Leon Gordenker (Princeton University)

39 Preventive Human Rights Strategies (2010)
by Bertrand G. Ramcharan (Geneva Graduate Institute of International and Development Studies)

38 African Economic Institutions (2010)
by Kwame Akonor (Seton Hall University)

37 Global Institutions and the HIV/AIDS Epidemic (2010)
Responding to an international crisis
by Franklyn Lisk (University of Warwick)

36 Regional Security (2010)
The capacity of international organizations
by Rodrigo Tavares (United Nations University)

35 The Organisation for Economic Co-operation and Development (2009)
by Richard Woodward (University of Hull)

34 Transnational Organized Crime (2009)
by Frank Madsen (University of Cambridge)

33 The United Nations and Human Rights (2nd edition, 2009)
A guide for a new era
by Julie A. Mertus (American University)

32 The International Organization for Standardization (2009)
Global governance through voluntary consensus
by Craig N. Murphy (Wellesley College) and
JoAnne Yates (Massachusetts Institute of Technology)

31 Shaping the Humanitarian World (2009)
by Peter Walker (Tufts University) and Daniel G. Maxwell (Tufts University)

30 Global Food and Agricultural Institutions (2009)
by John Shaw

29 Institutions of the Global South (2009)
by Jacqueline Anne Braveboy-Wagner (City College of New York, CUNY)

28 International Judicial Institutions (2009)
The architecture of international justice at home and abroad
by Richard J. Goldstone (Retired Justice of the Constitutional Court of South Africa) and Adam M. Smith (Harvard University)

27 The International Olympic Committee (2009)
The governance of the Olympic system
by Jean-Loup Chappelet (IDHEAP Swiss Graduate School of Public Administration) and Brenda Kübler-Mabbott

26 The World Health Organization (2009)
by Kelley Lee (London School of Hygiene and Tropical Medicine)

25 Internet Governance (2009)
The new frontier of global institutions
by John Mathiason (Syracuse University)

24 Institutions of the Asia-Pacific (2009)
ASEAN, APEC, and beyond
by Mark Beeson (University of Birmingham)

23 United Nations High Commissioner for Refugees (UNHCR) (2008)
The politics and practice of refugee protection into the twenty-first century
by Gil Loescher (University of Oxford), Alexander Betts (University of Oxford), and James Milner (University of Toronto)

22 Contemporary Human Rights Ideas (2008)
by Bertrand G. Ramcharan (Geneva Graduate Institute of International and Development Studies)

21 The World Bank (2008)
From reconstruction to development to equity
by Katherine Marshall (Georgetown University)

20 The European Union (2008)
by Clive Archer (Manchester Metropolitan University)

19 The African Union (2008)
Challenges of globalization, security, and governance
by Samuel M. Makinda (Murdoch University) and
F. Wafula Okumu (McMaster University)

18 Commonwealth (2008)
Inter- and non-state contributions to global governance
by Timothy M. Shaw (Royal Roads University)

17 The World Trade Organization (2007)
Law, economics, and politics
by Bernard M. Hoekman (World Bank) and
Petros C. Mavroidis (Columbia University)

16 A Crisis of Global Institutions? (2007)
Multilateralism and international security
by Edward Newman (University of Birmingham)

15 UN Conference on Trade and Development (2007)
by Ian Taylor (University of St. Andrews) and
Karen Smith (University of Stellenbosch)

14 The Organization for Security and
Co-operation in Europe (2007)
by David J. Galbreath (University of Aberdeen)

13 The International Committee of the Red Cross (2007)
A neutral humanitarian actor
by David P. Forsythe (University of Nebraska) and
Barbara Ann Rieffer-Flanagan (Central Washington University)

12 The World Economic Forum (2007)
A multi-stakeholder approach to global governance
by Geoffrey Allen Pigman (Bennington College)

11 The Group of 7/8 (2007)
by Hugo Dobson (University of Sheffield)

10 The International Monetary Fund (2007)
Politics of conditional lending
by James Raymond Vreeland (Georgetown University)

9 The North Atlantic Treaty Organization (2007)
The enduring alliance
by Julian Lindley-French (Center for Applied Policy, University of Munich)

8 The World Intellectual Property Organization (2006)
Resurgence and the development agenda
by Chris May (University of the West of England)

7 The UN Security Council (2006)
Practice and promise
by Edward C. Luck (Columbia University)

6 Global Environmental Institutions (2006)
by Elizabeth R. DeSombre (Wellesley College)

5 Internal Displacement (2006)
Conceptualization and its consequences
by Thomas G. Weiss (The CUNY Graduate Center) and David A. Korn

4 The UN General Assembly (2005)
by M. J. Peterson (University of Massachusetts, Amherst)

3 United Nations Global Conferences (2005)
by Michael G. Schechter (Michigan State University)

2 The UN Secretary-General and Secretariat (2005)
by Leon Gordenker (Princeton University)

1 The United Nations and Human Rights (2005)
A guide for a new era
by Julie A. Mertus (American University)

Books currently under contract include:

The Regional Development Banks
Lending with a regional flavor
by Jonathan R. Strand (University of Nevada)

Millennium Development Goals (MDGs)
For a people-centered development agenda?
by Sakiko Fukada-Parr (The New School)

The Bank for International Settlements
The politics of global financial supervision in the age of high finance
by Kevin Ozgercin (SUNY College at Old Westbury)

International Migration
by Khalid Koser (Geneva Centre for Security Policy)

The International Monetary Fund (2nd edition)
Politics of conditional lending
by James Raymond Vreeland (Georgetown University)

The UN Global Compact
by Catia Gregoratti (Lund University)

Institutions for Women's Rights
by Charlotte Patton (York College, CUNY) and Carolyn Stephenson (University of Hawaii)

International Aid
by Paul Mosley (University of Sheffield)

Coping with Nuclear Weapons
by W. Pal Sidhu

Global Governance and China
The dragon's learning curve
edited by Scott Kennedy (Indiana University)

The Politics of Global Economic Surveillance
by Martin S. Edwards (Seton Hall University)

Mercy and Mercenaries
Humanitarian agencies and private security companies
by Peter Hoffman

Regional Organizations in the Middle East
by James Worrall (University of Leeds)

Reforming the UN Development System
The Politics of Incrementalism
by Silke Weinlich (Duisburg-Essen University)

The International Criminal Court
The Politics and practice of prosecuting atrocity crimes
by Martin Mennecke (University of Copenhagen)

BRICS
by João Pontes Nogueira (Catholic University, Rio de Janeiro) and Monica Herz (Catholic University, Rio de Janeiro)

The European Union (2nd edition)
Clive Archer (Manchester Metropolitan University)

Protecting the Internally Displaced
Rhetoric and reality
Phil Orchard (University of Queensland)

The Arctic Council
Governance within the Far North
by Douglas C. Nord (University of Umeå)

For further information regarding the series, please contact:

Nicola Parkin, Editor, Politics & International Studies
Taylor & Francis
2 Park Square, Milton Park, Abingdon
Oxford OX14 4RN, UK
Nicola.parkin@tandf.co.uk
www.routledge.com